T.D. Walshaw

THE I.S.O. SYSTEM OF UNITS

An Introduction

ARGUS BOOKS

Argus Books
Argus House
Boundary Way
Hemel Hempstead
Herts HP2 7ST
England

First published by Argus Books 1991

ISBN 1 85486 063 1

Phototypesetting by Photoprint, Torquay, Devon
Printed and bound in Great Britain by Biddles Ltd, Guildford
and King's Lynn

Contents

List of Tables

Preface

When this book was first suggested to me, I had my doubts. Doubts as to whether it was necessary, for one thing, and whether I should attempt to write it, for another. After all, it is something over 70 years since I 'learned my tables' and I had been using feet and inches, pounds and ounces, and gallons and pints exclusively for the first 50 years of that period. (To say nothing of foot-pounds, British Thermal Units, and tons per square inch!) As to the necessity, I felt that the younger generation (anyone less than 40!) who had been educated into the Metric system did not need it, while on the other hand, most of the older generation would never change. (That is what they told me, at any rate!)

However, second thoughts suggested that there *was* something useful to be said. First, it was clear to me that talk of 'Metrication' missed the point, suggesting as it did that the change involved no more than using millimetres in place of inches – no great matter; I have machines calibrated in both systems and happily swap from one to the other. The fact that the 'Systéme Internationale' (S.I.) had brought humanity, for the very first time, a totally coherent system of measuring *everything* seemed to escape attention completely. (The S.I. system has affected the Metric countries as much as ourselves, as I explain later.) Secondly, even in the limited field of linear dimensions, there were still problems on both sides. Much of the industrial world still uses inches, pounds and gallons, so that it is not only the 'generation gap' which stands between the use of S.I. or Imperial units, but also the standards of measurements of whole nations.

So, I accepted the commission, and have tried to explain not only the difference between the systems, but also *why* the change was found to be necessary. I shall also try to show that the S.I. system does have considerable advantages. At the same time, I have tried to avoid the trap which so many apologists have fallen into – of simply quoting 'conversion factors' which are easy to use. I have assumed that you will prefer to have these factors presented as accurately as is possible, so that you can round them off to suit your own purposes. At the same time, I have worked out 'ready reckoner' tables for the more frequently needed units which will enable you to read off figures from either system into the other.

This book is designed to serve all who are concerned with measurements; those brought up in traditionally Metric countries, who have not only to come to terms with S.I. but also with drawings etc from 'Imperial' countries; those who, like me, come to the new system very late in life; and, I hope, to those in countries who have (so far) no intention of changing from the Imperial system. Whether I have succeeded in my task remains to be seen!

TDW

Introduction

It is unfortunate that the introduction of the I.S.O. (or (S.I.) system of measurements was publicised as 'Metrication', and perhaps even more unfortunate that the main reasons given for the change were the need to conform to the practice in the EEC. Both were true, of course, but miss the main point of the exercise. As a result, firstly the many and considerable advantages of the I.S.O. system were never properly appreciated and, secondly, more than a little hostility was generated to what was regarded as a 'foreign' system of weights and measures. The fact that all the Metric countries – far wider than Europe – had to face similar changes was not appreciated, either. True, they did not have the enormous expense of renewing measuring equipment and altering machine tools, but they were, without exception, forced to make fundamental alterations to their methods of dealing with force, energy, and work, and saw their beloved 'centimetre' relegated to no better than a 'permitted' dimension! Why, then, so great an unheaval, affecting as it has done, all fields of technological and scientific endeavour?

The units we (the Imperial countries) use have evolved over the millennia. The Mile came to use from the Romans – 1000 paces, each of two strides. The Furlong is Anglo-Saxon, the distance the team could pull a plough without a breather, and the Acre the area the team could cover in a day. In the feudal system of agriculture the land was divided into strips, and the width of an acre having a length of one furrow was divided into four – the Rod, Pole or Perch. Our Foot seems to have come from the East – it is half a Cubit (or one of them, for there were many) and was divided into 12 to make the Inch (or 'digit') Why 12? Because it was convenient to the technology of the day. The inch also, was divided into 12 Lines and, when fine mechanisms appeared (e.g. clocks) the line also was divided into 12. Cloth was measured by the Nail, (about 2½ inches) 16 of which made a Yard and 20 to the Ell. The nail and ell have disappeared, but the Yard, standardised by Henry VIII, is the basic Imperial unit for all linear measurements, large or small.

Similarly, units of 'capacity' or volume grew up according to usage; Gills, Pints, Quarts, Gallons, Pecks and Bushels, (to say nothing of Firkins, Ankers, Chaldrons and Weys!), none of which had any connection at all with length measurement. Weights, too, were developed each to serve their office – Grains, Ounces and Pounds at one end of the scale, Hundredweights and Tons at the other. (And, to add a little fun to the arithmetic at school, 'Troy' ounces were not the same weight as those used in buying sweets!)

The development of Engineering first and, later, Science, demanded the use of compound units – Footpounds, Pounds per square inch, British Thermal Units, and so on. Each of these needed yet another constant of subdivision – foot-pounds to inch-ounces, for example. It says much for the flexibility of the human mind that generations of students coped with all this – and generations of **Amateur** engineers, laymen, did so too. But there was one further issue which was more serious.

The distinction between 'Mass' and 'Weight' was not realised by the ancients. Even when the **effect** of mass was appreciated, e.g. when starting off a heavy waggon, the effect was put down to the **weight** of the load. However, as Engineering developed and dynamic loading became important, so did this distinction. To make this clearer, 'Mass' can be regarded as that property of a body which resists acceleration; 'Inertia', if you like. 'Weight' is no more than the *force* exerted on a mass by the pull of gravity. The weight of a body varies, even on the surface of the earth, but its mass remains constant, even in outer space.

Unfortunately, although the early engineers and scientists realised the distinction, they used the same **name** for the two properties; a mass of one pound had a weight of one pound, and so on. This meant that in order to calculate, say, the acceleration forces due to the mass of an engine piston its weight was divided by 'g' – the acceleration due to gravity – to find its mass. This not only introduced yet another 'conversion constant' but could lead to mistakes.

The system had outgrown its usefulness. Very convenient indeed in a feudal agricultural economy, and serving the purposes of commerce very well (a dozen has many advantages over a ten!) but it simply did not match up to the needs of Science and Engineering.

The continental Metric system was conceived in France at the end of the eighteenth century and formally adopted at the beginning of the 19th, to cope with a similar but more complex problem. The French faced not only a plethora of their own units, but also the fact that, for example, the length of the foot varied from state to state – or even from city to city; 11.14 English inches in Dresden, 11.28 in Hamburg, 11.72 in Rome, 12.45 in Vienna and 13.68 in Venice There were similar differences in the other units as well, of course – some very considerable.

The replacement system was, first, **Decimal**, in that all multiples of units were related by factors of ten. It was **Homonominal**, the multiples and submultiples all carrying the same *name*, but distinguished by decimal prefixes – Kilo-, Centi- etc. And it was, to some extent, **Rational**, as there was a definite relationship between all the basic units, Metre, Gram, Litre, etc. However, there was still no distinction between Mass and Weight, and the Gramme and Kilogramme were used for both, still requiring the use of 1/g in dynamic calculations. In some scientific circles there **was** an attempt to overcome this in the Centimetre–Gram–Second system; a new unit of 'Force', the Dyne, was introduced This was the force needed to accelerate one gram at a rate of 1 centimetre/second/second; with a corresponding unit of energy, the Erg. However, these were far too small – one pound-force = 445 000 Dyne, approximately!

The common engineering metric system was the Metre–Kilogram–Second (MKS) system, where the Kg was used both for mass and force. However, even after a century of development the Metric system had a number of defects quite apart from the Mass–Weight confusion. Derived units of (e.g.) work and energy involved conversion factors as in the Imperial system and there was a curious anomaly in that the definition of the Litre made it not **quite** 1000 cubic centimetres! The metric system, although it got over the problems facing Napoleon as he moved from country to country, did not properly serve the needs of the late twentieth century. It is as well to remember that 'Weights and Measures' span the **whole** of human activity; some are concerned with the wavelength of light while others are attempting to measure the width of the galaxy; the mass of a single proton matters to some while others consider 1000 tons a moderate weight. The field is enormous, and the desirability of a uniform and **consistent** system became apparent many decades ago.

Thus the I.S.O. (International Standards Organisation) system was born, involving representatives of the National Standards Institutions of almost every country in the world. It enables *all* the units for *any* branch of Science or Engineering to be defined in terms of no more than seven basic units. The unit of force is no longer tied to the acceleration due to gravity and there is no distinction between the various forms of energy, all – electrical, mechanical, thermal, etc – are expressed in the same units. Apart from the units of time and angle (hours, minutes seconds, and degrees or radians) the system is **decimal** but, with very few exceptions, **Kilodecimal**, having intervals of 1000 in place of ten. As will be seen later, this means that in all but a very few cases there are no 'constants of conversion' (such as those needed to convert BThU to Ft.Lbf) in the I.S.O. system; calculation is restricted to the arithmetic of the quantities only.

The fundamental units are, of course, **Length, Mass and Time**, being the METRE, KILOGRAM (not the Gram) and SECOND. All electrical units are derived from the AMPERE, which is itself related to the fundamentals. The S.I. system works on the concept of **Absolute Temperature**, measured in DEGREES KELVIN, but this involves no change in the magnitude of the degree, for $1°K = 1°C$. There is a change of name, however, the degree **Centigrade** being renamed the degree **Celsius**. The other two basic units are the CANDELA, which relates to illumination, and the MOLE, in the fields of Chemistry and Physics. Neither are new, but both have been redefined. There are also two *secondary* units, the RADIAN for plane angle and the STERADIAN, for solid angle, both of which have been in common use for a long time.

From these basic units are derived a new unit of FORCE, the Newton, which is quite independent of gravity, the definition of which is dealt with on page 21. Finally, ENERGY, no matter in what form, is measured in JOULES; again, a long established unit, but given a new importance. The definitions of these basic, and of the derived and compound units, (such as pressure, stress, density, etc) are given later, but a Joule is one Newton.metre, and 1 Joule/sec. is equal to one Watt. The *basic* definitions are rather abstruse and of no importance in service. The weights and measures which we use in practice all refer to standard masses and length bars held in the various National Standards Offices. However, the definitions of *derived* units should be studied with care.

The end result is the 'Systéme Internationale' (S.I.), a system of measurement the units of which are consistent in *all* fields of Science and Engineering. Almost all measurements can be made in units built up from the basic ones of Mass, Length and Time with no 'conversion factors'. Concepts of force and energy are clarified and computation, whether manual or by machine, is simplified. (I give examples later of the last point.)

It should be emphasised that no-one would claim that the S.I. system is more **'Scientific'** than its predecessors. It is not – for all systems are quite arbitrary and the Cubits, Spans and Digits used in building the Ark were just as appropriate to the 'science' of Noah's time as the Yard was to Henry VIII and the Metre is today. But S.I. does meet the needs of twentieth century Science and Engineering in a way which neither the Imperial nor the old Metric system ever could.

That there are some imperfections cannot be denied. The physical magnitude of the unit of 'Pressure' – Force per unit Area – is an inconvenience, and there are those who hold that the unit of length could have been more happily chosen, and that the I.S.O. should have started from scratch and devised a totally new system to avoid these minor disadvantages. This could have been done, of course, but critics should remember that the third fundamental unit of measurement is **TIME**, and to have altered the lengths of Seconds, Minutes and Hours to make them consistent would have been a formidable task – even if to do so were acceptable to the world at large! Compared with the manifest advantages, the very few weaknesses of the system are of negligible importance. The main difficulty is that associated with change of any sort, and I deal with that later. Meanwhile, to illustrate the difference between working in Imperial and S.I. units I give the following examples.

(1a) *Imperial Units*

On test an experimental hot-air engine holds a force of 4.3 ounce on a dynamometer arm 3.5 inches long at a speed of 1200 rpm. Calculate the output in Watts.

Work done/minute =

$$\frac{4.3}{16} \times \frac{3.5}{12} \times 2 \times \pi \times 1200 \text{ Ft.Lbf/min.}$$

$$= \underline{591} \text{ Ft.Lbf/min.}$$

1 Watt = 44.23 Ft.Lbf.

Hence output $= \dfrac{591}{44.23}$

$$= \textbf{\underline{13.36 watt}}$$

(1b) *S.I. Units*

The corresponding load is 1.195 Newton, the arm 89mm (0.089m) and the speed 20 revs/sec.

Work done/second =
$$1.195 \times 0.089 \times 2 \times \pi \times 20 \text{ Newton.metre/sec}$$
$$= 13.36 \text{ Nm/sec.}$$
$$(1 \text{ Nm/sec} = 1 \text{ J/sec} = 1 \text{ watt})$$

Output   = **<u>13.36 Watt</u>**

(2a) *Imperial Units*

A works reservoir draws from a catchment area of 13 sq.miles with an average rainfall of 31 inches/year, and with a 'yield ratio' of 0.38. What is the reliable yield in millions of gallons/day (MGD)? [1 Cu.ft = 6.23 gal.]

Bring all to foot units.

Rainfall	= 31/12 = 2.5833 ft.
Area	= 5280^2 = 362.42 × 10^6 sq.ft.
Volume	= 362.42 × 2.5833 × 10^6 Cu.ft/yr
	= 936.24 × 6.23 × 10^6 gall/yr.
	= 15.98 MGD
<u>Safe yield</u>	= 0.38 × 15.98 = **6.07 MGD**

(2b) *S.I. Units*

The area is now 33.67 sq. Km, the rainfall is 787 mm (0.787 metre) and the yield ratio 0.38 as before. In S.I. units the yield is stated in Megalitre/day. (1 cu.metre = 1000 Litre)

Volume/year	$= 33.67 \times 10^6 \times 0.787 \times 10^3$
	$= 26.498 \times 10^9$ Litre/year.
	$= 72.60 \times 10^6$ Litre/**day**
Safe yield	$= 72.60 \times 0.38$
	= **27.587 ML/day**

The work in the last (2b) case could have been written in a single line if the yield ratio had been included in the initial calculation.

A further example (given for a different purpose) shown on page 76 will further emphasise the contrast between working in Imperial and in S.I. units. Once they come to terms with the problem of 'scale', and become accustomed to the units, few users regret the change.

Prefixes of Magnitude

Whereas the former Imperial units of convenience operated on a mixture of binary, decimal and duodecimal multipliers each with a separate name (inch, foot, ounce, pound, etc), the original Metric system retained the same unit name throughout and multiples were wholly decimal, indicated by a prefix of magnitude – millimetre, Kilogram, Hectolitre, etc. The S.I. uses a similar system, but is **kilodecimal**, in that the general practice is to use multiples and submultiples which change by a factor of 1000 rather than 10. This reduces the risk of confusion (e.g. between 'deci' and 'Deka'), as well as the number of words to be remembered by lay users. It also takes account of the type of scaling used in many measuring instruments. There are a very few exceptions to the rule, chiefly for the convenience in particular trades and professions. The S.I. prefixes are shown overleaf, the 'non-preferred' in italics.

Table 1.

Prefix	Abbreviation	Multiplier	
Peta	P	10^{15}	1 000 000 000 000 000
Tera	T	10^{12}	1 000 000 000 000
Giga	G	10^{9}	1 000 000 000
Mega	M	10^{6}	1 000 000
Kilo	K	10^{3}	1000
Hecto	*H*	10^{2}	*100*
Deka	*da*	*10*	*10*
None	**–**	**Unit**	**1**
deci	*d*	10^{-1}	*0.1*
centi	*c*	10^{-2}	*0.01*
milli	m	10^{-3}	0.001
micro**	μ	10^{-6}	0.000 001
nano	n	10^{-9}	0.000 000 001
pico	p	10^{-12}	0.000 000 000 001
femto	f	10^{-15}	0.000 000 000 000 001

Note** The word **'micron'** is to be used in place of 'micro-metre' (μm) to avoid confusion with the measuring instrument.
'centimetre' is a permitted unit in the retail trade in cloth etc.
'Hectare' is the **approved** unit for land measurement. (1Ha = 100are = 0.01Km²)
'Centistoke' and **'Centi**poise' are permitted units of viscosity.

USAGE

In use the prefix forms part of the unit described and there should be no stop between them; thus, mm or Kg, not m.m or K.g. Any applied power relates to the unit as a whole, including the prefix, so that 3Km² is 3 square kilometres – 3 (1000m)² – not 3000 sq.metre.

In speech the prefix and unit are rendered in full, not as abbreviations – e.g. 'Kilogram', not 'kay-gee-em'. In which connection note that 'kilometre' is properly pronounced '**Kil**oh-**mee**ter' and **NOT** 'killo**omm**itter', the latter pronunciation being that of a measuring instrument.

Where basic units are multiplied or divided – e.g. moment of inertia in kilogram metre squared – the 'point' between the two units is optional but if no point is used a gap should be left – i.e. Kg.m^2 or Kg m^2. But when divided the fraction bar is always used – e.g. Kg/m^3 for mass per unit volume.

Note that the abbreviation of the unit is always a lower case letter (thus 'm' for metre, not 'M') **except** where the unit is named after a person when capital letters are used; e.g. 'A' for Ampere, 'J' for Joule. Capital letters are used as abbreviations for prefixes *greater* than 10, lower case for those less than 10 – e.g. mm for millimetre, but Km for kilometre.

EXCEPTION. Because a lower case letter 1 can be confused with the figure 'one' it is recommended that the litre be denoted by capital 'L'.

Care must be taken in writing certain combinations. Thus it is wiser to write 0.003 N rather than 3 mN, as the latter could be read as a metre-Newton instead of milli-Newton.

SPECIAL CASE. Prefixes of units of MASS

The unit of Mass is the Kilogram, which bears the prefix 'Kilo' already. 1/1000 Kg is the GRAM, not the 'Milli-kilogram'; and 1/1000 000 Kg is a MILLIGRAM, not a μ-Kilogram. Similarly with multiples. 1000 Kg is 1 Mg, (or, colloquially, 1 TONNE) and NOT a Kilo-Kilogram. In other words, though the Kilogram is the S.I. unit, all prefixes are applied to the sub-unit, the gram.

Definitions and Derivations of Units

BASIC UNITS

I.S.O. (International Organisation for Standardisation) has endorsed the S.I. (Système Internationale) units of measurement in terms of seven basic units and two supplementaries. All measurements in any technology can be made in units derived from these and all are consistent one with another. These units are defined as follows.

METRE The basic unit of length, and is 1 650 763.73 wavelengths of the radiation from *Krypton–86* under certain specified conditions. This is an **absolute** dimension depending on a physical phenomenon.

KILOGRAM is the basic unit of mass, being the mass of the 'Prototype Kilogram', held at the International Bureau of Weights and Measures. This is an **arbitrary** standard, not derived from any physical quantity, so that reference back to the prototype mass (which is a block of platinum-iridium alloy) is necessary, but the prototype mass approximates very closely indeed to that of 0.001 cubic metres of pure water at its maximum density condition.

SECOND The second is the S.I. unit of time interval measurement, being the duration of 9 192 631 770 cycles of the radiation from Caesium-133 under specified conditions. This is an **absolute** dimension.

AMPERE is the unit of electric current, being that constant (direct) current which will produce a force of

2×10^7 Newton/metre of length when flowing in two straight parallel conductors set one metre apart and having infinite length and negligible cross-section. This definition is derived from the mechanical S.I. units of Mass and Length. (The 'Newton' is a derived S.I. unit of **force** – see later.)

KELVIN is the S.I. unit of temperature, one degree Kelvin ($1°K$) being defined as $1/(273.16)$ of the thermo-dynamic (i.e. 'absolute') temperature of the triple point of water. The $°K$ is identical with the Celsius (formerly 'Centigrade') degree – $°C$, and $0°C = 273.15°K$. The $°K$ depends on a **physical** phenomenon.

CANDELA is the unit of luminous intensity, defined in terms of the luminous intensity of a true 'black body' of area $1/600\,000$ sq.metre under certain specified conditions of temperature and pressure.

MOLE The Mole (mol) is the S.I. unit of the **amount** of a substance, and is defined as that amount which contains the same number of elementary 'entities' as there are atoms in $0.012Kg$ of Carbon–12. When used the 'entities' must be specified – e.g. atoms, molecules, etc.

RADIAN is the S.I. unit of plane angle. This is a 'supplementary' unit, defined as the plane angle between two radii of a circle which cut an arc of the circle exactly equal in length to the radius. Thus there are 2π radians to the circle.

STERADIAN is the (supplementary) unit of solid angle, defined as the solid angle which, with its vertex at the centre of a sphere, cuts off an area at the surface equal to that of a square of side equal to the radius of the sphere.

It should be noted that these definitions are intended only to **establish** the magnitude of the units. The methods used for their practical determination may differ e.g. the definition of the **Ampere** is no more than a special case of the general formulae covering the forces between current carrying conductors. Again, for practical purposes, the various National Standards Offices hold copies of the standard prototype metre, against which the standards used by manufacturers of measuring equipment can be checked.

SECONDARY & DERIVED UNITS.

Area The S.I. unit is the **square metre (m^2)** and its multiples and submultiples – e.g. Km^2 and mm^2. For land measurement the **Hectare** (0.01 Km^2) may be used.

Volume and Capacity The S.I. unit is the **cubic metre (m^3)** and its multiples and submultiples – e.g. mm^3 etc. The **Litre** (0.001 m^3) has been adopted for the measurement of fluids. NOTE that the litre has been redefined so that 1 mL $= 1$ cm^3 *exactly and the mL should now be used in place of the cc or cm^3*

Velocity The unit of velocity or speed is the **metre/sec (m/s or $m.sec^{-1}$)** with variations formed by using multiples of the metre. For certain purposes the **kilometre/hour (Km/h)** may be used, and at sea the International Knot ($1852m/h$) is approved.

Angular Velocity The S.I. unit is the **Radian per second (rad/s)**, but revolutions per second or per minute (rev/s or rev/min) are also approved. For very low velocities degrees/min (°/min) or revolutions/hour (rev/h) may be used.

Acceleration The S.I. unit is the **metre/second/second (m/sec^2)**. In Geodesy the **Galileo (Gal) $= 0.01m/sec^2$** is used. The standard 'acceleration due to gravity', or 'g' is 9.80665 m/sec^2.

Force The S.I. unit is the **Kilogram metre/second²**, and has a special name, the NEWTON. It is defined as *That force which when applied to a mass of one kilogram gives it an acceleration of one metre/sec².* The symbol is 'N', and is subject to the usual kilodecimal multipliers as required – mN, KN, etc. (Note that a capital N is used.) To give some idea of scale in other systems, 1 Lb.force is about 4.45N, 1 Kg.force (kilopond) is about 9.8N, and 1 Dyne = 10^{-5}N or 10μN exactly.

Weight 'Weight' is really a force, being that exerted on a mass by gravitational attraction. For this reason, even on Earth, the 'weight' of an object can vary from place to place. In the UK the Weights and Measures Act allows 'weight' to be used to describe a 'mass' in commercial transactions so that for these purposes only the unit of weight is the Kilogram. As most weighing devices are, in fact, mass comparators, and the standard Kg. 'weight' is that of the I.S.O. Kg mass × g (9.80665) there will be no error. However, laymen should realise that it is of paramount importance to distinguish between 'mass' and 'weight' when engaged in engineering or scientific work. For example, a load which 'weighs' 100 on a machine calibrated in kilogram will exert a **force** on its foundation supports of 100g Newton = 980.1 N. It is wisest to avoid the term 'weight' except when buying coal or groceries, and use the term 'mass', even when that mass is determined by 'weighing' it on a pair of scales or a spring balance.

Pressure and Stress The unit is the NEWTON/SQ.METRE (N/m^2) which is given the special name of PASCAL (Pa). When using N/m^2 either or both elements of the unit can carry the prefix of magnitude e.g. MN/m^2, N/mm^2 (both of which are equal to 1 MPa) or in extreme cases, Kn/mm^2 = GPa.

The Pascal is a very small unit (1 Pa = 0.000 14 Lbf/sq.in Approx) so that the MPa has become the usual form. It is in order to use the compound form – MN/m^2 etc – if more convenient in the calculation.

An approved **non-S.I.** unit is the BAR = 0.1 MPa. This is commonly used for steam pressure etc (but never for 'stress'), 1 Bar being approximately atmospheric pressure.

Pressure may be expressed in terms of the height of a column of liquid; if so, the units should be metres or millimetres.

Energy, Work, Heat, etc The unit of energy, in any of its forms, is the JOULE, defined as the amount of work done when a force of 1 Newton is displaced a distance of 1 metre in the direction of application of the force. The Joule is compatible with all forms of energy, 1 J = 1 Newton.Metre = 1 Watt.Second = 1 Pascal.metre^3.

Heating values of fuels are stated in MJ/m^3, MJ/Kg or MJ/Litre, and the Enthalpy of steam usually in KJ/Kg. **Specific Heat** is quoted in KJ/Kg.°K. (1 BThU = 1055J approx; 1 International KgCal = **4.1868** J exactly.)

Power The unit of power in all its forms is the WATT. 1W = 1 Joule/second. The usual multiples are applies – KW, GW, etc.

Moments are obtained in the usual way, by multiplying the appropriate S.I. units – Nm, etc. NOTE The Imperial system distinguishes between 'work' (or 'energy') and 'moments of forces' by writing 'foot pounds' (ft.lbf) for the former and 'pounds feet' for the latter. The S.I. system makes no such distinction, with the result that (eg) a bending moment *could* be expressed in Joules! In places where such confusion could arise it is recommended that moments be written as **metre Newton**. However, this must always be abbreviated as m.N to avoid confusion with the milli Newton (mN).

Moment of Inertia is expressed in Kg.m^2.

Section Modulus and **Second Moment of Area** are expressed in the normal Metric units of m^3 and m^4 respectively.

Momentum Linear momentum is expressed in Kilogram-Metre/second (Kg.m./sec) and Angular momentum in Kg.m^2/sec. Multipliers should be applied with some care; 1 Kg.m^2/sec = 10^6 Kg.mm^2/sec = 10^3 gram.mm^2/sec.

Heat Flux Heat flow rate is the same as power and is measured in **Watts**, and Heat Flux Density (i.e. heat flow/unit area) in Watts/m^2. **Thermal conductivity** is quoted in Watts/metre.°Kelvin (W/m°K) Note that this is dimensionally identical to the conventional concept of heat units per unit time per unit area through unit thickness per degree temperature difference. The Watt includes the time element (1 watt = 1 J/sec) Note that it implies a thickness measured in **metres** and the appropriate multiplier must be used when measuring in mm.

Electrical Units These are all derived from the fundamental definition of the **Ampere**, see p.18. The unit of E.M.F. ('tension') is the VOLT, defined as the difference in potential between two points of a conductor carrying a current of one Ampere when the power dissipated between the two points is one Watt. The OHM (Ω) is the unit of Resistance, defined as the electrical resistance between two points of a conductor when a constant potential difference between these points of one Volt produces a current of one Ampere. The COULOMB (C) is the unit of **Electrical Quantity**, being that quantity carried by a current of one Ampere in one second. Electrical **Capacitance** is measured in FARADS (F) (in practice, usually microfarads – μF). This is defined as the capacitance of a capacitor in which a potential difference of one volt between the plates appears when charged with one Coulomb of electricity. (i.e. 1 Farad = 1 Ampere-second per volt). The HENRY (H) is the unit of **Inductance**, defined as the inductance of a closed circuit in which a potential (EMF) of one Volt is produced when the current is reduced at a uniform rate of one Ampere/second. (1H = 1 Volt-second/ampere)

Magnetic Flux This is measured in WEBERS (WB), one

Weber being the magnetic flux which, linking a circuit of one turn, would produce an EMF of one volt if reduced to zero in one second at a uniform rate. **Flux Density** is measured in TESLAS (T), one Tesla being one Weber per square meter of magnetic circuit area.

MISCELLANEOUS

The measure of Frequency remains unchanged in definition, being number of cycles per second, but is renamed the HERZ (H). Low frequencies should, however, be stated in terms of the actual phenomena and time – e.g. vibrations/min or periods/hour.

Viscosity Dynamic viscosity is measured in Newton seconds/sq.metre, or Pascal-seconds. The unit for Kinematic Viscosity is the square metre/second. Both bear decimal relationships with the former units of POISE and STOKE, as 1 Pa.sec = 10 Poise, and 1 m^2/sec = 10^4 Stokes.

Density is measured in whatever combinations of mass and volume, area, or length may be convenient. E.g. Kg/m for barstock etc, Kg/m^2 for sheet, and Kg/m^3 or g/mm^3 for solids. The density of liquids is usually indicated in Kg/Litre. **Specific Volume** is the inverse of density, and is expressed in cubic metres/kilogram or m^3/Kg.

Mass Flow is normally stated in Kg/sec, but Kg/hr is also approved. **Volume flow** may be stated either in m^3/sec or L/sec. In the water industry, flow in rivers is measured in 'CUMECS' = cu.m/sec, in pipework in litres/sec, and in aqueducts in Megalitre/day.

Impact Resistance (Izod, Charpy, etc tests) These numbers represent *energy absorption* stated in Ft.Lbf. or m.Kgf, and under I.S.O. definitions ought to be quoted in *Joules*. It is, however, permissible and perhaps more convenient to use the Newton metre.

Conversion Between Systems

Conversion between S.I., CGS, MKS and Imperial systems can be calculated from the *defined* equations for the basic units, viz:

1 Lb = 0.453 592 37 Kg
1 Yd = 0.9144 Metre. (1 In = 25.4mm)

These are *exact*. In addition, the following energy relationships have been defined exactly:

1 IT Kg Cal = 4186.8 Nm (Joule)
1 BThU/Lb   = 2.326 KJ/Kg

In the tables which follow (*pp. 27 to 37*) figures shown in **BOLD** type are 'exact'; others have been calculated to the number of significant figures shown using a 10-digit calculator. In some cases the value of 'g' is needed. The mean terrestial value has been standardised at **9.806 65 m/Sec²** or 32.1740 Ft/sec². Conversion from degrees to radians involves the factor PI, which is still subject to refinement by computer addicts. 22/7 is accurate only to 3 decimal places; **355/113** will serve to six places (see p. 32).

IMPORTANT NOTE. Many of the factors and tabulated values will be found to differ from those given in published works and reference books, especially if printed before the passing of the Weights & Measures Act of 1963. Quite apart from the redefinition of the yard (above) the **Litre** has also been redefined, and is now **exactly 0.001** cubic metre. (The previous definition was in terms of the volume of a standard mass of water.)

The conversion *factors* are stated to a degree of

precision which is quite unnecessary for most engineering purposes. For convenience 'direct conversion' tables for the most commonly used units are given on pp. 38 to 67. These have been calculated from the basic units and rounded off to one or two more decimal places than really necessary. Most of these tabulated figures will be accurate to better than 0.05%.

Conversion of **dimensions** on engineering drawings is dealt with in detail on page 77. When such conversion is needed to accommodate a single component – e.g. for a metric ball bearing and an imperial shaft and housing – the conversion must be as near as possible. In general, however, the recommendation is that the whole drawing should be redimensioned, using limits and tolerances appropriate to the work in hand. It is quite wrong to state a 'rule dimension' to several decimal places, whether I.S.O. at 20.6375mm or Imperial at 0.8125!

ACCURATE CONVERSION FACTORS

LENGTH

	Metre	Inch	Foot	Yard	Chain	Furlong	Mile
Metre =	1	39.370 079 7	3.280 839 9	1.093 613 3	0.049 710	0.004 971	0.6214×10^{-3}
Inch =	0.02540	1	0.083 33...	0.0277̇	0.001 263	–	–
Foot =	0.30480	12	1	0.3333̇	0.015 15̇15̇	0.001 51̇5̇	0.1894×10^{-3}
Yard =	0.91440	36	3	1	0.045 45̇45̇	0.004 54̇5̇	0.56818×10^{-3}
Chain =	20.1168	792	66	22	1	0.1000	0.012 50
Furlong =	201.1680	7920	660	220	10	1	0.1250
Mile =	1609.3440	63 360	5280	1760	80	8	1

1/1000 inch = 0.0254 mm   1/100 mm = 0.000 393 7 inch.
1 Angstrom Unit (Å) = 10^{-10} metre. 1 Light-year = 9.4605×10^{15} metre. 1 Parsec = 3.085×10^{16} metre = 3.261 Light-year.
1 Rod, Pole or Perch = 5½ Yd = 5.0292m. 1 Ligne = 1/12 in. = 2.1167 mm. 1 Hand = 4 in. = 101.6mm.

AREA

1 square yard = **0.836 127 360** Square metres. 1 square metre = **10.763 911 110 56** square feet.
1 Hectare = 10 000 square metres. 1 square kilometre = 100 Hectare.

	mm^2	m^2	In^2	Ft^2	Yd^2
1 mm² =	**1**	**0.000 001**	0.001 550	10.8 × 10⁻⁶	1.196 × 10⁻⁶
1 m² =	**1 000 000**	**1**	**1550 003**	10.763 910	1.195 990 1
1 in² =	**645.160**	**0.000 645 16**	**1**	0.006 944...	0.000 772
1 Ft² =	929.03 × 10⁻⁶	**0.092 903 04**	**144**	**1**	0.111 111..
1 Yd² =	83.613 × 10⁶	**0.836 127 36**	**1296**	**9**	**1**

	Ha	$Sq.Km$	$Acre$	$Sq.Mile$
1 Ha =	**1**	**0.01**	**2.471 053 8**	0.003 861
1 Km² =	**100**	**1**	**247.105**	0.386 102
1 Acre =	0.404 686	0.004 047	**1**	**0.001 562 5**
1 Sq.Mi =	258.999	2.589 99	**640**	**1**

1 Rood = 1210 Yd² = 0.25 Acre = 1447.15 m²

VOLUME

1 Cubic Yard = **0.764 554 875 984** m³. 1 Cu.Metre = **35.314 666 88** Cu.Foot. 1 Litre = **0.001** Cubic Metre.

		mL = cm³	Litre	Cu. Metre	Cu. Inch.	Cu. Foot	Cu. Yard
1 mL	=	**1**	**0.001**	**0.000 001**	**0.061 023 74**	–	0.001 130 8
1 Litre	=	**1000**	**1**	**0.001**	**61.023 74**	0.035 315	**1.307 950 7**
1 Metre³	=	**10⁶**	**1000**	**1**	**61.023 74**	**35.314 670**	0.000 021 4
1 Cu. Inch	=	16.387 064	0.016 387	0.000 016	**1**	0.5787×10^{-3}	0.037 037 04
1 Cu. Foot	=	28 316.84	**28.316 84**	**0.028 316 846**	**1728**	**1**	**1**
1 Cu. Yard	=	764 555.2	764.555 2	0.764 555	**46 656**	**27**	

		Litre	Cu.Metre	U.K.Gall	U.S.Gall
1 Litre	=	**1**	**0.001**	**0.219 969**	0.264 174
1 Cu.M.	=	**1000**	**1**	**219.969**	264.172
1 UK Gal	=	4.546 095	0.004 546	**1**	1.200 95
1 US Gal	=	3.78541	0.003 785	**0.832 674**	**1**

(1 US gall = 231 cu.ins, 1 UK gall = 277.420 032 Cu. Ins)

1 Fl. Ounce (= 8 drachms) = 28.4131 mL. 1 Gall = 8 Pint. 1 Pint = 20 Fl.oz. (Liquid Measure) 1 Bushel = 4 Peck.
1 Peck = 2 Gal solid measure.

MASS

1 Lb. av (Imperial) = 0.453 592 37 Kg. 1Kg = 2.204 622 622 Lb.Av. 1 Grain = 0.064 798 91 gram.

	Gram	Kilogram	Grain	Ounce	Pound	Slug
1 Gram =	1	0.001	15.43236	0.035 274	0.002 204 62	0.014 593
1 Kg =	1000	1	15 432.36	35.27337	2.204 622 622	14.5939
1 Grain =	0.064 798 91	—	1	0.002 285 71	0.000 142 9	0.000 004 4
1 Ounce =	28.349 523	0.028 35	437.5	1	0.062 5	0.000 1 943
1 Pound =	453.592 37	0.453 592 37	7000	16	1	0.031 081
1 Slug =	14 593.9	14.593 9	2.252×10^5	514.784	32.17400	1

(The **SLUG** is an Imperial unit of mass used chiefly in the aerospace industry. One Slug is that mass which, if operated upon by a force of one lbf, accelerates at 1 foot/sec².)

	Kilogram	Tonne	Cwt	Ton (UK)	Ton (US)
1 Kg =	1	0.001	0.019 684 15	0.000 984 21	0.001 102 31
1 Tonne =	1000	1	19.684 15	0.984 207	1.102 311
1 Cwt =	50.802 3	0.058 023	1	0.05	0.056
1 Ton =	1016.046 9	1.016 047	20	1	1.120
1 Ton US =	907.185	0.907 185	17.857 143	0.982 857 14	1

1 Ounce TROY measure = **480** Grains = **31.1035** gram. 1 CARAT = **200mg**
1 STONE = 14 Lb. 1 QUARTER = **28** Lb. 1 CENTAL = **100** Lb. 1 QUINTAL = **100** Kg.

DENSITIES Etc.

Mass/Unit Length

1 Lb/Inch	=	**17.8580 Kg/m**	1 Kg/m	=	**0.055 977 4 Lb/in.**
1 Lb/Foot	=	**1.488 164 Kg/m**	1 Kg/m	=	**0.671 968 9 Lb/Ft**
1 Lb/Yard	=	**0.496 055 Kg/m**	1 Kg/m	=	**2.015 91 Lb/Yd**

Mass/Unit Area

1 Lb/Ft^2	=	4.882 428 Kg/m^2	1 Kg/M^2	=	0.204 816 Lb/ft^2
1 Oz/Yd^2	=	0.033 906 Kg/m^2	1 Kg/m^2	=	29.494 Oz/Yd^2
1 Cwt/Acre	=	125.535 Kg/Ha	1 Kg/m^2	=	8 921.8 Lb/Acre

1 Tonne/Hectare = 8 cwt/Acre approx.

Mass/Unit Volume

1 Lb/Cu.In	=	27 679.90 Kg/m^3	1 Kg/m^3	=	3.6127×10^{-6} Lb/In^3
1 Lb/Cu.Ft	=	16.0185 Kg/m^3	1 Kg/m^3	=	62.4278×10^{-3} Lb/Ft^3
1 Lb/Gal	=	99.773 Kg/m^3	1 Kg/m^3	=	10.023×10^{-3} Lb/Gal
1 Ton/Cu.Yd	=	1.329 Tonne/m^3	1 T/m^3	=	0.75245 Ton/Cu Yd.

1 Gram/mL = 1 Gram/cc = 1000Kg/m^3. 1 gram/Litre = 1 Kg/m^3

Specific Volume

1 Cu.Ft/Lb = **0.062 427 9 m^3/Kg.** 1 m^3/Kg = 16.018 48 Cu.Ft/lb.

SPEED or VELOCITY

1 Ft/Sec	=	**0.3048 m/sec**	1 m/sec	=	**3.280 839 9 Ft/sec**
1 Ft/min	=	**0.005 08 m/sec**	1 m/sec	=	196.85 Ft/min
	=	**0.3048 m/min**	1 m/sec	=	**39.370 0787 In/Sec**
1 In/sec	=	**0.0254 m/sec**	1 m/sec	=	2.236 932 mph .
1 mile/hr	=	**0.447 04 m/sec**	1 Km/hr	=	0.6214 mph
	=	1.6093 Km/hr			

ACCELERATION

1 Ft/sec^2	=	**0.3048 m/sec^2**	1m/sec^2	=	**3.280 839 9 Ft/sec^2**

ANGULAR VELOCITY and ACCELERATION

The basic units are the same in all systems – i.e. rads/sec or rads/sec^2. Conversion from or to 'revolutions' involves the factor PI (1 rev = 2π radians) which is indeterminate. In ascending order of accuracy PI may be taken as:

22/7 = 3.14286; 3.142; 3.14159; 355/113 = 3.141 592 9;
   or 3.141 592 65 4.

1 rad/sec = 9.549 297 rpm.   1 rpm = 0.104 719 8 rad/sec.

FORCE

		Newton	Kg.Force	Poundal	Lb.Force	Ton Force
1 N	=	**1**	**0.101 971 62**	7.233 01	**0.224 808 94**	$1.003\ 61 \times 10^{-4}$
1 Kgf	=	**9.806 65**	**1**	70.931 444	**2.204 62**	$9.842\ 07 \times 10^{-4}$
1 Pdl	=	**0.138 255 17**	0.014 098	**1**	**0.031 081 0**	$1.387\ 54 \times 10^{-5}$
1 Lbf	=	**4.448 221 62**	0.453 592 37	**32.1740**	**1**	$4.464\ 29 \times 10^{-4}$
1 Tonf	=	9964.02	1016.047	**72 069.76**	**2240.0**	**1**

1 DYNE = 10^{-5} NEWTON. 1 NEWTON = 0.1 Megadyne.

Note The DYNE is that force required to accelerate 1 gram at 1 cm/sec^2
The POUNDAL is that force required to accelerate 1 lb.mass at 1 ft/sec^2
The KgForce is that force required to accelerate 1 Kg at "standard gravity", 9.80 665 m/sec^2
The LbForce is that force required to accelerate 1 Lb at "standard gravity", 32.1740 ft/sec^2

FORCE/UNIT AREA – PRESSURE and STRESS

	N/m²(Pa)	Lbf/Ft²	Lbf/In²	Kgf-cm²	Bar	Tonf/In²
N/m²(Pa) =	1	0.020 885 4	$1.450\ 377\ 4 \times 10^{-4}$	$1.019\ 716\ 2 \times 10^{-6}$	10^{-5}	6.4749×10^{-8}
Lbf/Ft² =	47.880 258	1	$6.944\ 44 \times 10^{-3}$	4.88243×10^{-5}	$4.788\ 03 \times 10^{-4}$	3.1×10^{-6}
Lbf/In² =	6894.744 8	0.00694...	1	0.070 307	0.068 947	4.4643×10^{-4}
Kgf/cm² =	98.0665×10^{3}	2048.165	14.223 369	1	0.980 665	6.3947×10^{-3}
BAR =	10^{5}	2088.547	14.503 774	1.019 716 2	1	6.4749×10^{-3}
Tonf/In² =	$1.544\ 423 \times 10^{7}$	3.23×10^{5}	2240	157.487 53	154.443	1

Note. The Pascal (Pa) is so small that MPa (MN/m² = N/mm²) may be regarded as the basic "Unit".
The BAR is commonly used for steam and gas pressure, and N/mm² (or MPa) for stresses in metals.

Other Units. 1 mm Hg = **1.333 22 millibar** = **133.322 Pa.** 1 Inch W.G. = **249.089 Pa. 1 In.Hg** = 3386.39 Pa
1 Poundal/in² = **214.295 04 Pa.** 1 Pa = 0.004 666 Pdl/In²

MOMENTS OF FORCE and TORQUE, BENDING MOMENT, Etc.

1 N.m = 0.737 564 5 Lbf.Ft. = 0.101 971 6 Kgf.m
1 Lbf.Ft = 1.355 813 6 N.m   1 Kgf.m. = **9.806 65 N.m.**

ENERGY

All forms are expressed in **Joules**. **1 Joule** = **1 N.m** = **10^7 Erg** = **9.478 69 × 10^{-4} BThU** = **1 Watt.Second.**

	Joule	KWHr	Ft.Lbf	HP.Hr	Kgf.m	BThU
1 Joule =	**1**	2.778×10^{-7}	0.737 562	$3.725\ 06 \times 10^{-7}$	0.101 972	9.478 **17 × 10^{-4}**
1 KWHr =	**3.6 × 10^6**	**1**	**2 654 155.5**	1.3405	0.3671×10^6	**3413**
1 Ft.Lbf =	**1.355 818 2**	37.68×10^{-6}	**1**	50.51×10^{-6}	0.138 255	**1.285 × 10^{-3}**
1 HP.Hr =	**2.684 52 × 10^6**	**0.7457**	**1.98 × 10^6**	**1**	$0.273\ 75 \times 10^6$	**2545**
1 BThU =	**1055.056 044**	2.931×10^{-4}	778.3	0.3931×10^{-3}	107.59	**1**
1 Kgf.m =	**9.806 6**	2.724×10^{-6}	7.233	3.653×10^{-6}	**1**	9.295×10^{-3}

1000 International Table Calorie = 4186.8 J = 3.968 32 BThU = 2.204 62 CHU = 3088 Ft.Lbf = 1 K.Cal$_{IT}$

POWER

1 KW = **1000 J/Sec** = 737.562 Ft.Lbf/Sec = 1.341 02 HP
1 Ft.Lbf/Sec = 1.355 82 Watt   1 HP = **550 Ft.Lbf/Sec**.

TEMPERATURE

$1°C = \dfrac{5}{9}°F.$ $1°F = 1.8°C$

$T°C = \dfrac{5}{9}(T°F - 32)$

$T°F = 1.8\,T°C + 32.$
Abs. °K = °C + 273.15
Abs. °F = °F + 459.69. (Known as Deg. Rahkine or °R)

ENTHALPY (HEAT CONTENT)

1 BThU/Lb = 2.326 KJ/Kg *by definition*
1 KJ/Kg = 0.429 922 6 BThU/Lb.
1 KJ/m^3 = 26.839 2 × 10^{-3} BThU/Ft^3
1 BThU/Ft^3 = 37,258.9 J/m^3 = 8.899 15 K.Cal$_{IT}$/m^3
1 Therm/Gal(Imp) = 23.208 MJ/Litre. (1 Therm = 100,000 BThU)

SPECIFIC HEAT

1 J/Kg°K) = 0.238 845 9 × 10^{-3} BThU/(Lb°F)
= 0.185 893 8 Ft.Lbf/(Lb.°F).
1 BThU/(Lb°F) = 4186.8 J/(Kg°K)
1 BThU/(Ft^3°F) = 67,066 J/(m^3°K) = 16.019 Cal$_{IT}$/(m^3°K)

COEFFICIENT OF HEAT TRANSFER

The S.I.Unit is the Watt/(metre.°K) = W/(m .°C)
1 BThU/(Ft.Hr.°F) = 1.731 W/(m .°K)
1 K.Cal(m .Hr.°K) = 1.163 W/(m .°K)

Note. This coefficient is a complex derived unit, being *heat units flowing in unit time over unit area for each unit temperature difference assuming unit thickness* and formerly conveniently stated as BThU/sq.ft/Hr/°F/ Ft.thickness, or KgCal/sq.metre/Hr/°C/metre thickness.

Care is needed, as coefficients may be found in some references which are not in consistent units – e.g. feet and inches, or metres and mm. In addition, the time factor may be quoted in hours or seconds. These inconsistencies must be remedied by converting the individual factors before using the overall conversion shown above.

FUEL CONSUMPTION Etc.

1 BThU/BHP.Hr = 1.4149 KJ/KWHr.
1 KJ/KWHr   = 0.7068 BThU/BHP.Hr.
1 Lb/BHP.Hr   = 453.6 gram/BHP/Hr = 608.28 gram/KWH
1 Gram/BHP.Hr = 0.0022 Lb/BHP.Hr.
1 Gram/KWH   = 0.00164 Lb/BHP.Hr.

$$\text{Litres/100 Km} = \frac{282}{\text{mpg}}. \qquad \text{Mpg} = \frac{282}{\text{L/100 Km}}$$

VISCOSITY

1 Poise = 0.1 N.Sec/m^2 = 0.1 Pa.Sec = 0.020 885 4 Lbf.Sec/Ft^2
1 Stoke = 10^{-4} m^2/Sec = 1.076 391 × 10^{-3} Ft^2/Sec
        = 0.155 00 Ins^2/Sec

ELECTRICAL UNITS

The only conversions likely to be needed are those involving **forces**, bearing in mind that most countries using Imperial units applied the MKS system for electrical work. Both the Dyne and the Kilopond (Kilogram-force) were used in this system. The Newton is the replacement. For other units, though some of the names are altered the change from MKS to S.I. units may involve no more than a factor of some power of 10.

WORKING CONVERSION TABLES

Tables 2 to 15 which follow present direct conversion of length, area, volume, mass, force, pressure or stress, and (for the convenience of motorists) fuel consumption, rounded to a reasonable degree of accuracy for practical purposes. Being decimal, the table values can be extended quite easily. Thus 900 sq.metres is found by multiplying the figure for 90 by ten; and 956 sq.m. is found by first reading 1022.6 × 10 = 10 226 sq.ft and then adding 64.56 making 10 290.56 sq.ft.

Table 2 Length
Inches and Millimetres
One Inch = 25.40 millimetres exactly. 1 mm = 0.039 370 078 inch.
Inch to mm, 0 – 59

Ins.		10	20	30	40	50
0		254·0	508·0	762·0	1016·0	1270·0
1	25·4	279·4	533·4	787·4	1041·4	1295·4
2	50·8	304·8	558·8	812·8	1066·8	1320·8
3	76·2	330·2	584·2	838·2	1092·2	1346·2
4	101·6	355·6	609·6	863·6	1117·6	1371·6
5	127·0	381·0	635·0	889·0	1143·0	1397·0
6	158·4	406·4	660·4	914·4	1168·4	1422·4
7	177·8	431·8	685·8	939·8	1193·8	1447·8
8	203·2	457·2	714·2	965·2	1219·2	1473·2
9	228·6	482·6	736·6	990·6	1244·6	1498·6

Decimal inch to mm.

Ins.	mm	Ins.	mm	Ins.	mm
·001	·0254	·01	0·254	·1	2·54
·002	·0508	·02	0·508	·2	5·08
·003	·0762	·03	0·762	·3	7·62
·004	·1016	·04	1·016	·4	10·16
·005	·1270	·05	1·270	·5	12·70
·006	·1524	·06	1·524	·6	15·24
·007	·1778	·07	1·778	·7	17·78
·008	·2032	·08	2·032	·8	20·32
·009	·2288	·09	2·286	·9	22·86

Table 2 contd.

Fractions of an inch to mm.

Inches.				mm	Inches.				mm
		1/64	·015625	·3969			33/64	·515625	13·0969
	1/32		·03125	·7937		17/32		·53125	13·4937
		3/64	·046875	1·1906			35/64	·546875	13·8906
1/16			·0625	1·5875	9/16			·5625	14·2875
		5/64	·078125	1·9844			37/64	·578125	14·6844
	3/32		·09375	2·3812		19/32		·59375	15·0812
		7/64	·109375	2·7781			39/64	·609375	15·4781
1/8			·125	3·1750	5/8			·625	15·8750
		9/64	·140625	3·5719			41/64	·640625	16·2719
	5/32		·15625	3·9687		21/32		·65625	16·6687
		11/64	·171875	4·3656			43/64	·671875	17·0656
3/16			·1875	4·7625	11/16			·6875	17·4625
		13/64	·203125	5·1594			45/64	·703125	17·8594
	7/32		·21875	5·5562		23/32		·71875	18·2562
		15/64	·234375	5·9531			47/64	·734375	18·6531
1/4			·25	6·3500	3/4			·75	19·0500
		17/64	·265625	6·7469			49/64	·765625	19·4469
	9/32		·28125	7·1437		25/32		·78125	19·8437
		19/64	·296875	7·5406			51/64	·796875	20·2406
5/16			·3125	7·9375	13/16			·8125	20·6375
		21/64	·328125	8·3344			53/64	·828125	21·0344
	11/32		·34375	8·7312		27/32		·84375	21·4312
		23/64	·359375	9·1281			55/64	·859375	21·8281
3/8			·375	9·5250	7/8			·875	22·2250
		25/64	·390625	9·9219			57/64	·890625	22·6219
	13/32		·40625	10·3187		29/32		·90625	23·0187
		27/64	·421875	10·7156			59/64	·921875	23·4156
7/16			·4375	11·1125	15/16			·9375	23·8125
		29/64	·453125	11·5094			61/64	·953125	24·2094
	15/32		·46875	11·9062		31/32		·96875	24·6062
		31/64	·484375	12·3031			63/64	·984375	25·0031
1/2			·5	12·7000					

Table 2 contd.

Millimetre to inch, 0 – 99

mm	10	20	30	40	50	60	70	80	90
0	·39370	·78740	1·18110	1·57480	1·96851	2·36221	2·75591	3·14961	3·54331
1 ·03937	·43307	·82677	1·22047	1·61417	2·00788	2·40158	2·79528	3·18898	3·58268
2 ·07874	·47244	·86614	1·25984	1·65354	2·04725	2·44095	2·83465	3·22835	3·62205
3 ·11811	·51181	·90551	1·29921	1·69291	2·08662	2·48032	2·87402	3·26772	3·66142
4 ·15748	·55118	·94488	1·33858	1·73228	2·12599	2·51969	2·91339	3·30709	3·70079
5 ·19685	·59055	·98425	1·37795	1·77165	2·16536	2·55906	2·95276	3·34646	3·74016
6 ·23622	·62992	1·02362	1·41732	1·81103	2·20473	2·59843	2·99213	3·38583	3·77953
7 ·27559	·66929	1·06299	1·45669	1·85040	2·24410	2·63780	3·03150	3·42520	3·81890
8 ·31496	·70866	1·10236	1·49606	1·88977	2·28347	2·67717	3·07087	3·46457	3·85827
9 ·35433	·74803	1·14173	1·53543	1·92914	2·32284	2·71654	3·11024	3·50394	3·89764

Millimetre to inch, 100 to 990

mm		100	200	300	400	500	600	700	800	900
0		3·93701	7·87402	11·8110	15·7480	19·6851	23·6221	27·5591	31·4961	35·4331
10	·39370	4·33071	8·26772	12·2047	16·1417	20·0788	24·0158	27·9528	31·8898	35·8268
20	·78740	4·72441	8·66142	12·5984	16·5354	20·4725	24·4095	28·3465	32·2835	36·2205
30	1·18110	5·11811	9·05513	12·9921	16·9291	20·8662	24·8032	28·7402	32·6772	36·6142
40	1·57480	5·51181	9·44883	13·3858	17·3228	21·2599	25·1969	29·1339	33·0709	37·0079
50	1·96851	5·90552	9·84252	13·7795	17·7165	21·6536	25·5906	29·5276	33·4646	37·4016
60	2·36221	6·29922	10·2362	14·1732	18·1103	22·0473	25·9843	29·9213	33·8583	37·7953
70	2·75591	6·69292	10·6299	14·5669	18·5040	22·4410	26·3780	30·3150	34·2520	38·1890
80	3·14961	7·08662	11·0236	14·9606	18·8977	22·8347	26·7717	30·7087	34·6457	38·5827
90	3·54331	7·48032	11·4173	15·3543	19·2914	23·2284	27·1654	31·1024	35·0394	38·9764

Decimal millimetre to inch.

mm	*Ins.*
0·001	·000039
0·002	·000079
0·003	·000118
0·004	·000157
0·005	·000197
0·006	·000236
0·007	·000276
0·008	·000315
0·009	·000354

mm	*Ins.*
0·01	·00039
0·02	·00079
0·03	·00118
0·04	·00157
0·05	·00197
0·06	·00236
0·07	·00276
0·08	·00315
0·09	·00354

mm	*Ins.*
0·1	·00394
0·2	·00787
0·3	·01181
0·4	·01575
0·5	·01969
0·6	·02362
0·7	·02756
0·8	·03150
0·9	·03543

Table 3 Area

Square inches/square millimetres. 1 Sq.In = **645.15** Sq.mm exactly.
Sq.ins to Sq.mm.

		1	2	3	4	5	6	7	8	9
0		645·16	1290·3	1935·5	2580·6	3225·8	3870·1	4516·1	5161·3	5806·4
0·1	64·52	709·68	1354·8	2000·0	2645·2	3290·3	3935·5	4580·6	5225·8	5871·0
0·2	129·03	774·19	1419·4	2064·5	2709·7	3354·8	4000·0	4645·2	5290·3	5935·5
0·3	193·55	838·71	1483·9	2129·0	2774·2	3419·3	4064·5	4709·7	5354·8	6000·0
0·4	258·06	903·22	1548·4	2193·5	2838·7	3483·9	4129·0	4774·2	5419·3	6064·5
0·5	322·58	967·74	1612·9	2258·0	2903·2	3548·4	4193·5	4838·7	5483·9	6129·0
0·6	387·10	1032·26	1677·4	2322·6	2967·7	3612·9	4258·1	4903·2	5548·4	6193·5
0·7	451·61	1096·77	1741·9	2387·1	3032·3	3677·4	4322·6	4967·7	5612·9	6258·1
0·8	516·13	1161·29	1806·4	2451·6	3096·8	3741·9	4387·1	5032·2	5677·4	6322·6
0·9	580·64	1225·80	1871·0	2516·1	3161·3	3806·4	4451·6	5096·8	5741·9	6387·1

Sq.mm to Sq.ins

		100	200	300	400	500	600	700	800	900
0		·1550	·3100	·4650	·6200	·7750	·9300	1·0850	1·2400	1·3950
10	·0155	·1705	·3255	·4805	·6355	·7905	·9455	1·1005	1·2555	1·4105
20	·0310	·1860	·3410	·4960	·6510	·8060	·9610	1·1160	1·2710	1·4260
30	·0465	·2015	·3565	·5115	·6665	·8215	·9765	1·1315	1·2865	1·4415
40	·0620	·2170	·3720	·5270	·6820	·8370	·9920	1·1470	1·3020	1·4570
50	·0775	·2325	·3875	·5425	·6975	·8525	1·0075	1·1625	1·3175	1·4725
60	·0930	·2480	·4030	·5580	·7130	·8680	1·0230	1·1780	1·3330	1·4880
70	·1085	·2635	·4185	·5735	·7285	·8835	1·0385	1·1935	1·3485	1·5035
80	·1240	·2790	·4340	·5890	·7440	·8990	1·0540	1·2090	1·3640	1·5190
90	·1395	·2945	·4495	·6045	·7595	·9145	1·0695	1·2245	1·3795	1·5345

Table 4 Area

Square Feet/Square metres. 1 Sq.Metre = 10.7639 Sq.Ft.
Sq.Ft to Sq.Metre

	0	10	20	30	40	50	60	70	80	90
0		·929	1·8581	2·7871	3·7161	4·6452	5·5742	6·5032	7·4322	8·3613
1	·0929	1·022	1·951	2·880	3·809	4·738	5·667	6·596	7·525	8·454
2	·1858	1·115	2·044	2·973	3·902	4·831	5·760	6·689	7·618	8·547
3	·2787	1·208	2·137	3·066	3·995	4·924	5·853	6·782	7·711	8·640
4	·3716	1·301	2·230	3·159	4·088	5·017	5·946	6·875	7·804	8·733
5	·4645	1·394	2·323	3·252	4·181	5·110	6·039	6·968	7·897	8·826
6	·5574	1·486	2·415	3·345	4·274	5·203	6·132	7·061	7·990	8·919
7	·6503	1·579	2·508	3·437	4·366	5·295	6·225	7·154	8·083	9·012
8	·7432	1·672	2·601	3·530	4·459	5·388	6·317	7·246	8·175	9·105
9	·8361	1·765	2·694	3·623	4·552	5·481	6·410	7·339	8·268	9·197

Sq. Metre to Sq.Ft

		10	20	30	40	50	60	70	80	90
0		107·64	215·28	322·92	430·56	538·20	645·83	753·47	861·11	968·75
1	10·764	118·40	226·04	333·68	441·32	548·96	656·60	764·24	871·88	979·51
2	21·528	129·17	236·80	344·44	452·08	559·72	667·36	775·00	882·64	990·28
3	32·292	139·93	247·57	355·21	462·85	570·49	678·13	785·76	893·40	1001·0
4	43·056	150·69	258·33	365·97	473·61	581·25	688·89	796·53	904·17	1011·8
5	53·820	161·46	269·10	376·74	484·38	592·01	699·65	807·29	914·93	1022·6
6	64·583	172·22	279·86	387·50	495·14	602·78	710·42	818·06	925·70	1033·3
7	75·347	182·99	290·63	398·26	505·90	613·54	721·18	828·82	936·46	1044·1
8	86·111	193·75	301·39	409·03	516·67	624·31	731·94	839·58	947·22	1054·9
9	96·875	204·51	312·15	419·79	527·43	635·07	742·71	850·35	957·99	1065·6

Table 5 Volume

Cubic Feet/Cubic Metres

1 Cu. Metre = 35.3147 Cu.Ft.
1 Cu. Ft = 0.028 316 Cu. Metre
(1 Cu.Metre = 1000 Litre)

Cubic Feet to Cubic Metre.

		10	20	30	40	50	60	70	80	90
0		·2832	·5663	·8495	1·1327	1·416	1·699	1·982	2·265	2·549
1	·0283	·3115	·5947	·8772	1·1610	1·444	1·727	2·010	2·294	2·577
2	·0566	·3398	·6230	·9061	1·1893	1·472	1·756	2·039	2·322	2·605
3	·0849	3681	·6513	·9345	1·2176	1·501	1·784	2·067	2·350	2·633
4	·1133	·3964	·6796	·9628	1·2459	1·529	1·812	2·095	2·379	2·662
5	·1416	·4248	·7079	·9911	1·2743	1·557	1·841	2·124	2·407	2·690
6	·1699	·4531	·7362	1·0194	1·3026	1·586	1·869	2·152	2·435	2·718
7	·1982	·4814	·7646	1·0477	1·3309	1·614	1·897	2·180	2·464	2·747
8	·2265	·5097	·7929	1·0760	1·3592	1·642	1·926	2·209	2·492	2·775
9	·2549	·5380	·8212	1·1044	1·3875	1·671	1·954	2·237	2·520	2·803

Cubic Metre to Cubic Feet

		10	20	30	40	50	60	70	80	90
0		353·15	706·29	1059·4	1412·6	1765·7	2118·9	2472·0	2825·2	3178·3
1	35·31	388·5	741·6	1095	1448	1801	2154	2507	2860	3214
2	70·63	423·8	776·9	1130	1483	1836	2190	2543	2896	3249
3	105·9	459·1	812·2	1165	1519	1872	2225	2578	2931	3284
4	141·3	494·4	847·6	1201	1554	1907	2260	2613	2966	3320
5	176·6	529·7	882·9	1236	1589	1942	2295	2649	3002	3355
6	211·9	565·0	918·2	1271	1624	1978	2331	2684	3037	3390
7	247·2	600·3	953·5	1307	1660	2013	2366	2719	3072	3426
8	282·5	635·7	988·8	1342	1695	2048	2401	2755	3108	3461
9	317·8	671·0	1024·1	1377	1730	2084	2437	2790	3143	3496

Table 6 Volume

Cubic Inches/Cubic Centimetres.

1 Cu.In = **16.327 064** cc (exactly)
1 Cu.Cm = 0.061 024 Cu.In.
(1 Litre = 1000 cc)

Cu. Ins to Cu.Cm.

		1	2	3	4	5	6	7	8	9
0		16·382	32·774	49·161	65·548	81·935	98·322	114·71	131·10	147·48
·1	1·639	18·026	34·413	50·800	57·187	83·574	99·961	116·35	132·74	149·12
·2	3·277	19·665	36·052	52·439	68·826	85·213	101·60	117·99	134·37	150·76
·3	4·916	21·303	37·690	54·077	70·464	86·952	103·24	119·63	136·01	152·40
·4	6·555	22·942	39·329	55·716	72·103	88·490	104·88	121·26	137·65	154·04
·5	8·194	24·581	40·978	57·355	73·742	90·129	106·52	122·90	139·29	155·68
·6	9·832	26·219	42·606	58·993	75·381	91·768	108·15	124·54	140·93	157·32
·7	11·471	27·858	44·245	60·632	77·019	93·406	109·79	126·18	142·57	158·95
·8	13·110	29·497	45·884	62·271	78·658	95·045	111·43	127·82	144·21	160·59
·9	14·748	31·135	47·523	63·910	80·297	96·684	113·07	129·46	145·84	162·23

Cu. Cm to Cu. Ins

		10	20	30	40	50	60	70	80	90
0		·610	1·220	1·831	2·441	3·051	3·661	4·272	4·882	5·492
1	·061	·671	1·281	1·892	2·502	3·112	3·722	4·333	4·943	5·553
2	·122	·732	1·343	1·953	2·563	3·173	3·783	4·394	5·004	5·614
3	·183	·793	1·404	2·014	2·624	3·234	3·844	4·455	5·065	5·675
4	·244	·854	1·465	2·075	2·685	3·295	3·906	4·516	5·126	5·736
5	·305	·915	1·525	2·136	2·746	3·356	3·967	4·577	5·187	5·797
6	·366	·976	1·587	2·197	2·807	3·417	4·028	4·638	5·248	5·858
7	·427	1·037	1·648	2·258	2·868	3·478	4·089	4·699	5·309	5·919
8	·488	1·098	1·709	2·319	2·929	3·539	4·150	4·760	5·370	5·980
9	·549	1·159	1·770	2·380	2·990	3·600	4·211	4·821	5·431	6·041

Table 7 Volume

Gallons (UK) / Litres

1 UK Gallon = 4.5461 Litre
1 Litre = 0.219 97 UK Gallon

UK Gallons to Litre

	0	10	20	30	40	50	60	70	80	90
0		45·46	90·92	136·4	181·8	227·3	272·8	318·2	363·7	409·1
1	4·55	50·01	95·47	140·9	186·4	231·9	277·3	322·8	368·2	413·7
2	9·09	54·55	100·0	145·5	190·9	236·4	281·9	327·3	372·8	418·2
3	13·64	59·10	104·6	150·0	195·5	240·9	286·4	331·9	377·3	422·8
4	18·18	63·65	109·1	154·6	200·0	245·5	290·9	336·4	381·9	427·3
5	22·73	68·19	113·7	159·1	204·6	250·0	295·5	340·9	386·4	431·9
6	27·28	72·74	118·2	163·7	209·1	254·6	300·0	345·5	390·9	436·4
7	31·82	77·28	122·7	168·2	213·7	259·1	304·6	350·0	395·5	440·9
8	36·37	81·83	127·3	172·8	218·2	263·7	309·1	354·6	400·0	445·5
9	40·91	86·38	131·8	177·3	222·8	268·2	313·7	359·1	404·6	450·0

Litres to UK Gallons

		10	20	30	40	50	60	70	80	90
0		2·20	4·40	6·60	8·80	11·00	13·20	15·40	17·60	19·80
1	0·22	2·42	4·62	6·82	9·02	11·22	13·42	15·62	17·82	20·02
2	0·44	2·64	4·84	7·04	9·24	11·44	13·64	15·84	18·04	20·24
3	0·66	2·86	5·06	7·26	9·46	11·66	13·86	16·06	18·26	20·46
4	0·88	3·08	5·28	7·48	9·68	11·88	14·08	16·28	18·48	20·68
5	1·10	3·30	5·50	7·70	9·90	12·10	14·30	16·50	18·70	20·90
6	1·32	3·52	5·72	7·92	10·12	12·32	14·52	16·72	18·92	21·12
7	1·54	3·74	5·94	8·14	10·34	12·54	14·74	16·94	19·14	21·34
8	1·76	3·96	6·16	8·36	10·56	12·76	14·96	17·16	19·36	21·56
9	1·98	4·18	6·38	8·58	10·78	12·98	15·18	17·38	19·58	21·78

Table 8 Mass

1 Pound = **0.453 592 37** Kilogram (Exact)
1 Kilogram = 2.204 622 Pound

Pound / Kilogram

Pound to Kilogram

		10	20	30	40	50	60	70	80	90
0		4·5359	9·072	13·608	18·144	22·680	27·216	31·752	36·287	40·823
1	·45359	4·9895	9·525	14·061	18·597	23·133	27·669	32·205	36·741	41·277
2	·90719	5·4431	9·979	14·515	19·051	23·587	28·123	32·659	37·195	41·731
3	1·36078	5·8967	10·433	14·969	19·504	24·040	28·576	33·112	37·648	42·184
4	1·81437	6·3503	10·886	15·422	19·958	24·494	29·030	33·566	38·102	42·638
5	2·26796	6·8039	11·340	15·876	20·412	24·948	29·484	34·019	38·555	43·091
6	2·72156	7·2575	11·793	16·329	20·865	25·401	29·937	34·473	39·009	43·545
7	3·17515	7·7111	12·247	16·783	21·319	25·855	30·391	34·927	39·463	43·998
8	3·62874	8·1647	12·701	17·237	21·772	26·308	30·844	35·380	39·916	44·452
9	4·08233	8·6183	13·154	17·690	22·226	26·762	31·298	35·834	40·370	44·906

Kilogram to Pound

		10	20	30	40	50	60	70	80	90
0		22·046	44·092	66·139	88·185	110·23	132·28	154·32	176·37	198·42
1	2·2046	24·251	46·297	68·343	90·389	112·44	134·48	156·53	178·57	200·62
2	4·4092	26·455	48·502	70·548	92·594	114·64	136·69	158·73	180·79	202·83
3	6·6139	28·660	50·706	72·752	94·799	116·84	138·89	160·94	182·98	205·03
4	8·8185	30·865	52·911	74·957	97·003	119·05	141·10	163·14	185·19	207·23
5	11·023	33·069	55·116	77·162	99·208	121·25	143·30	165·35	187·39	209·44
6	13·228	35·274	57·320	79·366	101·41	123·46	145·50	167·55	189·60	211·64
7	15·432	37·479	59·525	81·571	103·61	125·66	147·71	169·76	191·80	213·85
8	17·637	39·683	61·729	83·776	105·82	127·87	149·91	171·96	194·01	216·05
9	19·842	41·888	63·934	85·980	108·03	130·07	152·12	174·16	196·21	218·26

Table 9 Force

Imperial Pound – Force and S.I. Newton

1 Lbf = 4.448 Newton
1 Newton = 0.2248 Lbf

Lbf to Newton

		10	20	30	40	50	60	70	80	90
0		44·48	88·96	133·4	177·9	222·4	266·9	311·4	355·8	400·3
1	4·448	48·93	93·41	137·9	182·4	226·8	271·3	315·8	360·3	404·8
2	8·896	53·38	97·86	142·3	186·8	231·3	275·8	320·3	364·7	409·2
3	13·34	57·82	102·3	146·8	191·3	235·7	280·2	324·7	369·2	413·7
4	17·79	62·27	106·8	151·2	195·7	240·2	284·7	329·2	373·6	418·1
5	22·24	66·72	111·2	155·7	200·2	244·6	289·1	333·6	378·1	422·6
6	26·69	71·17	115·6	160·1	204·6	249·1	293·6	338·0	382·5	427·0
7	31·14	76·62	120·1	164·6	209·1	253·5	288·0	342·5	387·0	431·5
8	35·58	80·06	125·5	169·0	213·5	258·0	302·5	346·9	391·4	435·9
9	40·03	84·51	129·0	173·5	218·0	262·4	306·9	351·4	395·9	440·4

Newton to Lbf

		10	20	30	40	50	60	70	80	90
0		2·248	4·496	6·744	8·992	11·24	13·49	15·74	17·98	20·23
1	0·225	2·473	4·721	6·969	9·217	11·46	13·71	15·96	18·21	20·46
2	0·450	2·698	4·946	7·194	9·442	11·69	13·94	16·19	18·43	20·68
3	0·674	2·922	5·170	7·418	9·666	11·91	14·16	16·41	18·66	20·91
4	0·899	3·147	5·395	7·643	9·891	12·14	14·39	16·64	18·88	21·13
5	1·124	3·372	5·620	7·868	10·12	12·36	14·61	16·86	19·11	21·36
6	1·349	3·597	5·845	8·093	10·34	12·59	14·84	17·08	19·33	21·58
7	1·574	3·822	6·070	8·318	10·57	12·81	15·06	17·31	19·56	21·81
8	1·798	4·046	6·294	8·542	10·79	13·04	15·29	17·53	19·78	22·03
9	2·023	4·271	6·519	8·767	11·02	13·26	15·51	17·76	20·01	22·26

Imperial Ton-force and KiloNewton

Tonf to KN

Table 10 Force

1 Tonf = 9.963 KN
1 KN = 0.100(4) Tonf.

		10	20	30	40	50	60	70	80	90
0		99·6	199	299	399	498	598	697	797	897
1	9·96	109·6	209	309	409	508	608	707	807	907
2	19·93	119·6	219	319	418	518	618	717	817	917
3	29·89	129·5	229	329	428	528	628	727	827	927
4	39·86	139·5	239	339	438	538	638	737	837	937
5	49·82	149·5	249	349	448	548	648	747	847	947
6	59·78	159·4	259	359	458	558	658	757	857	957
7	69·74	169·4	269	369	468	568	668	767	867	967
8	79·71	179·4	279	379	478	578	678	777	877	976
9	89·68	189·3	289	389	488	588	688	787	887	986

KN to Tonf.

		10	20	30	40	50	60	70	80	90
0		1·004	2·01	3·01	4·02	5·02	6·02	7·03	8·03	9·04
1	·1004	1·104	2·11	3·11	4·12	5·12	6·12	7·13	8·13	9·14
2	·2008	1·204	2·21	3·21	4·22	5·22	6·22	7·23	8·23	9·24
3	·3012	1·305	2·31	3·31	4·32	5·32	6·32	7·33	8·33	9·34
4	·4016	1·405	2·41	3·41	4·42	5·42	6·42	7·43	8·43	9·44
5	·5020	1·505	2·51	3·51	4·52	5·52	6·52	7·53	8·53	9·54
6	·6024	1·605	2·61	3·61	4·62	5·62	6·62	7·63	8·63	9·64
7	·7028	1·706	2·71	3·71	4·72	5·72	6·72	7·73	8·73	9·74
8	·8032	1·806	2·81	3·81	4·82	5·82	6·82	7·83	8·83	9·84
9	·9036	1·907	2·91	3·91	4·92	5·92	6·92	7·93	8·93	9·94

Table 11 Force

Metric Kilogram-force to S.I.Newton

1 Kgf = 9.81 N
1 N = 0.1019 Kgf.

(The Kilogram-force is also known as the Kilopond).
Kgf to Newton

	0	10	20	30	40	50	60	70	80	90
0	0	98·1	196	294	392	490	588	687	785	883
1	9·81	107·9	206	304	402	500	598	697	794	892
2	19·61	117·7	216	314	412	510	608	706	804	902
3	29·42	127·5	226	324	422	520	618	716	814	912
4	39·23	137·3	235	333	431	530	628	726	824	922
5	49·03	147·1	245	343	441	539	637	736	834	932
6	58·84	156·9	255	353	451	549	647	745	843	941
7	68·65	166·7	265	363	461	559	657	755	853	951
8	78·45	176·5	275	373	471	569	667	765	863	961
9	88·26	186·3	284	382	481	579	677	775	873	971

Newton to Kgf.

		100	200	300	400	500	600	700	800	900
0		10·19	20·38	30·57	40·76	50·95	61·14	71·33	81·52	91·71
10	1·02	11·21	21·40	31·59	41·78	51·97	62·16	72·35	82·54	92·73
20	2·04	12·23	22·42	32·61	42·80	52·99	63·18	73·37	83·56	93·75
30	3·06	13·25	23·49	33·63	43·82	54·00	64·20	74·39	84·58	94·77
40	4·08	14·27	24·46	34·65	44·84	55·03	65·22	75·41	85·60	95·79
50	5·10	15·29	25·48	35·67	45·86	56·05	66·24	76·43	86·62	96·81
60	6·11	16·30	26·49	36·68	46·87	57·06	67·25	77·44	87·63	97·82
70	7·13	17·32	27·51	37·70	47·89	58·08	68·27	78·46	88·65	98·84
80	8·15	18·34	28·53	38·72	48·91	59·10	69·29	79·48	89·67	99·86
90	9·17	19·36	29·55	39·74	49·93	60·12	70·31	80·50	90·69	100·88

60

Table 12 Pressure

1 Lbf/In² = 0.006 89 MPa.
(1 MPa = 1N/mm²)

Lbf/Sq.In. and MegaPascal (MPa) or N/mm²

Lbf/In² to MPa

		10	20	30	40	50	60	70	80	90
0		·069	·138	·208	·276	·345	·413	·482	·551	·620
1	·007	·076	·145	·216	·282	·351	·420	·489	·558	·627
2	·014	·083	·152	·223	·289	·358	·427	·496	·565	·634
3	·021	·090	·158	·230	·306	·365	·434	·503	·572	·641
4	·028	·096	·165	·237	·313	·372	·441	·510	·579	·648
5	·035	·103	·172	·244	·310	·379	·448	·517	·586	·655
6	·041	·110	·179	·251	·317	·386	·455	·524	·593	·661
7	·048	·117	·186	·258	·324	·393	·462	·531	·599	·668
8	·055	·124	·193	·265	·330	·400	·468	·537	·606	·675
9	·062	·131	·200	·272	·337	·407	·475	·544	·613	·683

MPa to Lbf/in²

MPa	0	1	2	3	4	5	6	7	8	9
0		145	290	435	580	725	870	1015	1160	1305
0·1	14·5	160	305	450	595	740	885	1030	1175	1320
0·2	29·0	174	319	464	609	754	899	1044	1189	1334
0·3	43·5	189	334	479	624	769	914	1059	1204	1349
0·4	58·0	203	348	493	638	783	928	1073	1218	1363
0·5	72·5	218	363	508	653	798	943	1088	1233	1378
0·6	87·0	232	377	522	667	812	957	1102	1247	1392
0·7	101·5	247	392	537	682	827	972	1117	1262	1407
0·8	116·0	261	406	551	696	841	986	1131	1276	1421
0·9	130·5	276	420	566	711	856	1000	1146	1291	1434

Note. The difference between ABSOLUTE and GAUGE pressure is 0.101 MPa Approx.

Lbf/sq.in/Bar.

1 Bar = 10^5 Pascal = 14.503 77 Lbf/sq.in.
1 Lbf/sq.in = 0.068 947 Bar.

Table 13 Pressure

Lbf/Sq.In to Bar

	0	10	20	30	40	50	60	70	80	90
0		0·6895	1·3789	2·0684	2·7579	3·4473	4·1368	4·8263	5·5156	6·2052
1	0·0689	0·7584	1·4479	2·1373	2·8268	3·5613·	4·2057	4·8952	5·5847	6·2741
2	0·1379	0·8274	1·5168	2·2063	2·8958	3·5852	4·2747	4·9641	5·6536	6·3431
3	0·2068	0·8963	1·5858	2·2752	2·9647	3·6542	4·3436	5·0331	5·7226	6·4120
4	0·2759	0·9653	1·6547	2·3442	3·0336	3·7231	4·4126	5·1020	5·7915	6·4810
5	0·3447	1·0342	1·7237	2·4131	3·1026	3·7921	4·4815	5·1710	5·8605	6·5499
6	0·4137	1·1031	1·7926	2·4821	3·1715	3·8610	4·5505	5·2399	5·9294	6·6189
7	0·4826	1·1721	1·8616	2·5510	3·2405	3·9299	4·6194	5·3089	5·9983	6·6878
8	0·5516	1·2410	1·9305	2·6200	3·3094	3·9990	4·6884	5·3778	6·0675	6·7568
9	0·6205	1·310	1·9994	2·6889	3·3784	4·0678	4·7573	5·4468	6·1362	6·8257

Bar to Lbf/Sq.In.

	0	1·0	2·0	3·0	4·0	5·0	6·0	7·0	8·0	9·0
0		14·50	29·01	43·5	58·0	72·5	87·0	101·5	116·0	130·5
0·1	1·45	15·95	30·46	45·0	59·5	73·9	88·5	102·9	117·5	132·0
0·2	2·90	17·40	31·91	46·4	60·9	75·4	89·9	104·4	118·9	133·4
0·3	4·35	18·85	33·36	47·9	62·4	76·9	91·4	105·9	120·4	134·9
0·4	5·80	20·30	34·81	49·3	63·8	78·3	92·8	107·3	121·8	136·3
0·5	7·25	21·76	36·26	50·8	66·3	79·8	94·3	108·8	123·3	137·8
0·6	8·70	23·21	37·71	52·2	66·7	81·2	95·7	110·2	124·7	139·2
0·7	10·15	24·66	39·16	53·7	68·2	82·7	97·2	111·7	126·2	140·7
0·8	11·16	26·11	40·61	55·1	69·6	84·1	98·6	113·1	127·6	142·1
0·9	13·05	27·56	42·06	56·6	71·1	85·6	100·1	114·6	129·1	143·6

Table 14 Stress (Or High Pressure)

1 Tonf/sq.in = 15.444 MPa
1 MPa = 0.064 749 Tonf/Sq.In.

Tonf/Sq.In & MPa (N/mm²)

Tonf/Sq.In to MPa (N/mm²)

		10	20	30	40	50	60	70	80	90
0		154	309	463	618	772	927	1081	1235	1390
1	15·44	170	324	479	633	788	942	1097	1251	1405
2	30·89	185	340	494	649	803	958	1112	1266	1421
3	46·33	201	355	510	664	819	973	1127	1282	1436
4	61·78	216	371	525	670	834	988	1143	1297	1452
5	77·22	232	386	540	695	849	1003	1158	1313	1467
6	92·66	247	402	556	710	865	1019	1174	1328	1483
7	108·1	263	417	571	726	880	1034	1189	1344	1498
8	123·5	278	432	587	741	896	1050	1205	1359	1514
9	139·0	293	448	602	757	911	1066	1220	1375	1529

MPa to Tonf/Sq.In.

		10	20	30	40	50	60	70	80	90
0		·647	1·295	1·94	2·59	3·24	3·88	4·53	5·18	5·83
1	·065	·712	1·360	2·00	2·65	3·30	3·95	4·60	5·24	5·89
2	·129	·777	1·424	2·07	2·72	3·37	4·01	4·66	5·31	5·96
3	·194	·842	1·489	2·14	2·78	3·43	4·08	4·73	5·37	6·02
4	·259	·906	1·554	2·20	2·85	3·50	4·14	4·79	5·44	6·09
5	·324	·971	1·619	2·27	2·91	3·56	4·21	4·86	5·50	6·15
6	·388	1·036	1·683	2·33	2·98	3·63	4·27	4·92	5·57	6·22
7	·453	1·101	1·748	2·40	3·04	3·69	4·34	4·99	5·63	6·28
8	·518	1·165	1·813	2·46	3·11	3·76	4·40	5·05	5·70	6·35
9	·583	1·230	1·880	2·53	3·17	3·82	4·47	5·12	5·76	6·41

Table 15 Fuel Consumption

Miles/gallon to Litres/100 Km. and L/100 Km to MPG.

$$L/100 \text{ Km} = \frac{282}{MPG}$$

$$MPG = \frac{282}{L/100 \text{ Km}}$$

M.P.G. to Litres/100 Km

		10	20	30	40	50	60	70	80	90
0		28·2	14·1	9·4	7·1	5·6	4·70	4·03	3·53	3·13
1	282	25·7	13·4	9·1	6·9	5·5	4·62	3·97	3·48	3·10
2	141	23·5	12·8	8·8	6·7	5·4	4·55	3·92	3·44	3·07
3	94	21·7	12·3	8·5	6·6	5·3	4·48	3·86	3·40	3·03
4	70·5	20·1	11·8	8·3	6·4	5·2	4·4	3·81	3·36	3·00
5	56·4	18·8	11·3	8·1	6·3	5·1	4·34	3·76	3·32	2·97
6	47·0	17·6	10·8	7·8	6·1	5·0	4·27	3·71	3·28	2·94
7	40·3	16·6	10·4	7·6	6·0	4·9	4·21	3·66	3·24	2·91
8	35·3	15·7	10·1	7·4	5·9	4·86	4·15	3·62	3·20	2·88
9	31·3	14·8	9·7	7·2	5·8	4·8	4·09	3·57	3·17	2·85

Litres/100 Km to Miles/Gal.

	1	2	3	4	5	6	7	8	9	10	11	12
0	282	141	94	70·5	56·4	47·1	40·3	35·3	31·3	28·3	25·7	23·5
0·1	256	134	91	69	55	46·2	39·7	34·8	31·0	28·0	25·5	23·3
0·2	235	128	88	67	54	45·5	39·2	34·4	30·7	27·7	25·2	23·2
0·3	217	123	85	66	53	44·8	38·6	34·0	30·3	27·4	25·0	23·0
0·4	201	118	83	64	52	44·1	38·1	33·6	30·0	27·2	24·8	22·8
0·5	188	113	81	63	51	43·4	37·6	33·2	29·7	26·9	24·6	22·6
0·6	176	108	78	61	50	42·7	37·1	32·8	29·4	26·7	24·4	22·4
0·7	166	104	76	60	49	42·1	36·6	32·4	29·1	26·4	24·1	22·2
0·8	157	101	74	59	48·6	41·5	36·2	32·0	28·8	26·2	23·9	22·1
0·9	148	97	72	58	48	40·9	35·7	31·7	28·5	25·9	23·7	21·9

Table 16 Metric equivalent of Morse letter & number drills

Gauge No.	Decimal equivalent in.	ALTERNATIVE SIZES Rec. mm.	Exact mm.
80	0.013 5	0.35	0.34
79	0.014 5	0.38	0.37
78	0.016 0	0.40	0.41
77	0.018 0	0.45	0.46
76	0.020 0	0.50	0.51
75	0.021 0	0.52	0.53
74	0.022 5	0.58	0.57
73	0.024 0	0.60	0.61
72	0.025 0	0.65	0.64
71	0.026 0	0.65	0.66
70	0.028 0	0.70	0.71
69	0.029 2	0.75	0.74
68	0.031 0	0.80	0.79
67	0.032 0	0.82	0.81
66	0.033 0	0.85	0.84
65	0.035 0	0.90	0.89
64	0.036 0	0.92	0.91
63	0.037 0	0.95	0.94
62	0.038 0	0.98	0.97
61	0.039 0	1.00	0.99
60	0.040 0	1.00	1.02
59	0.041 0	1.05	1.04
58	0.042 0	1.05	1.07
57	0.043 0	1.10	1.09
56	0.046 0	1.20	1.18
55	0.052 0	1.20	1.32

Gauge No.	Decimal equivalent in.	ALTERNATIVE SIZES Rec. mm	Exact mm
53	0.059 5	1.50	1.51
52	0.063 5	1.60	1.61
51	0.067 0	1.70	1.70
50	0.070 0	1.80	1.78
49	0.073 0	1.85	1.85
48	0.076 0	1.95	1.93
47	0.078 5	2.00	1.99
46	0.081 0	2.05	2.06
45	0.082 0	2.10	2.08
44	0.086 0	2.20	2.18
43	0.089 0	2.25	2.26
42	0.093 0	2.40	2.37
41	0.096 0	2.45	2.44
40	0.098 0	2.50	2.49
39	0.099 5	2.55	2.53
38	0.101 5	2.60	2.58
37	0.104 5	2.65	2.64
36	0.106 5	2.70	2.71
35	0.110 0	2.80	2.79
34	0.111 0	2.80	2.82
33	0.113 0	2.85	2.87
32	0.116 0	2.95	2.95
31	0.120 0	3.00	3.05
30	0.128 5	3.30	3.26
29	0.136 0	3.50	3.45
28	0.140 5	3.60	3.57

Gauge No.	Decimal equivalent in.	ALTERNATIVE SIZES Rec. mm	Exact mm
26	0.147 0	3.70	3.73
25	0.149 5	3.80	3.80
24	0.152 0	3.90	3.86
23	0.154 0	3.90	3.91
22	0.157 0	4.00	3.99
21	0.159 0	4.00	4.04
20	0.161 5	4.10	4.09
19	0.166 0	4.20	4.22
18	0.169 0	4.30	4.30
17	0.173 0	4.40	4.39
16	0.177 0	4.50	4.50
15	0.180 0	4.60	4.57
14	0.182 0	4.60	4.62
13	0.185 0	4.70	4.70
12	0.189 0	4.80	4.80
11	0.191 0	4.90	4.85
10	0.193 5	4.90	4.92
9	0.196 0	5.00	4.98
8	0.199 0	5.10	5.06
7	0.201 0	5.10	5.11
6	0.204 0	5.20	5.18
5	0.205 5	5.20	5.22
4	0.209 0	5.30	5.31
3	0.213 0	5.40	5.41
2	0.221 0	5.60	5.61
1	0.228 0	5.80	5.79

Gauge Letter	Decimal equivalent in.	ALTERNATIVE SIZES Rec. mm	Exact mm
A	0.234 0	5.90	5.94
B	0.238 0	6.00	6.04
C	0.242 0	6.10	6.15
D	0.246 0	6.20	6.25
E	0.250 0	6.30	6.35
F	0.257 0	6.50	6.53
G	0.261 0	6.60	6.63
H	0.266 0	6.70	6.75
I	0.272 0	6.90	6.90
J	0.277 0	7.00	7.03
K	0.281 0	7.20	7.14
L	0.290 0	7.40	7.37
M	0.295 0	7.50	7.49
N	0.302 0	7.70	7.67
O	0.316 0	8.00	8.03
P	0.323 0	8.20	8.20
Q	0.332 0	8.40	8.43
R	0.339 0	8.60	8.61
S	0.348 0	8.80	8.84
T	0.358 0	9.10	9.09
U	0.368 0	9.30	9.34
V	0.377 0	9.60	9.58
W	0.386 0	9.80	9.80
X	0.397 0	10.10	10.08
Y	0.404 0	10.30	10.26
Z	0.413 0	10.50	10.49

Preferred Numbers

It will be found that many I.S.O. replacements of (e.g.) wire and sheet gauges, show what appear to be repetition of digits as the size increases. This is because such gauges have been taken to the nearest **preferred number**, drawn from a table of rational increases in value. The same applies to values of resistors and capacitors employed in electrical circuitry, limits and fits, preferred stock sizes, and, though not exact, screw thread diameters and pitch. These numbers go up in a geometric rather than a linear series from 10 to 100, smaller sizes being obtained by dividing and larger by multiplying by 10, 100 etc.

There are **five** such series, R5 where the increment is $^5\sqrt{10}$, approximately 60%, R10 at increment $^{10}\sqrt{10}$, about 25%; R20, $^{20}\sqrt{10}$, about 12%; R40 at $^{40}\sqrt{10}$, say 6%, and R80 at $^{80}\sqrt{10}$, approximately 3%. It will be found that R5 provides 5 numbers between 10 and 100, R10 offers 10, R20 gives 20 and so on. For general use the R10, R20 and R40 are usually employed, and these are found where 'First, Second and Third Preferences' are required. Rigid adherence to these series is not to be expected – e.g. copper tube at 0.7mm rather than 0.71mm thickness – but the general progression is and should be observed wherever possible. The table below gives the sequence of the R10, R20 and R40 series, each larger series incorporating the smaller, of course. The table has been expanded to cover sizes from 0.02 up to 25.0.

Table 17

Note that the progression of numbers runs **down** the columns.

Preference													
1st	2nd	3rd											
R10	R20	R40	·020	·040	·080	·160	·315	·63	1·25	2·5	5·0	10·0	20·0
		R40	·021	·042	·085	·170	·335	·67	1·32	2·65	5·3	10·6	21·2
	R20	R40	·022	·045	·090	·180	·355	·71	1·40	2·8	5·6	11·2	22·4
		R40	·024	·048	·095	·190	·375	·75	1·50	3·0	6·0	11·8	23·6
R10	R20	R40	·025	·050	·100	·200	·40	·80	1·60	3·15	6·3	12·5	25·0
		R40	·026	·053	·106	·212	·425	·85	1·70	3·35	6·7	13·2	
	R20	R40	·028	·056	·112	·224	·45	·90	1·80	3·55	7·1	14·0	
		R40	·030	·060	·118	·236	·475	·95	1·90	3·75	7·5	15·0	
R10	R20	R40	·032	·063	·125	·250	·50	1·00	2·00	4·0	8·0	16·0	
		R40	·034	·067	·132	·265	·53	1·06	2·12	4·25	8·5	17·0	
	R20	R40	·036	·071	·140	·28	·56	1·12	2·24	4·5	9·0	18·0	
		R40	·038	·075	·150	·30	·60	1·18	2·36	4·75	9·5	19·0	

NOTE. The series chosen as 'First' preference in any particular case will depend on the application and, of course, on the tolerances on adjacent values or sizes of component.

Using the I.S.O. Metric System

As in any other system it is important that the rules or 'grammar' be carefully observed, otherwise confusion is certain. Some of these rules are laid down – either in the British or the parent International Standard – while others have evolved in common usage. The more common are listed below.

(1) Except where the unit is named after a **person** (e.g. Ampere or Newton) the symbol for the unit itself is a lower case (small) letter. m for Metre, g for Gram, but A for Ampere, N for Newton, Pa for Pascal, K for Kelvin and so on. The **exception** is the Litre, always written as L, as l can read as 'one'.

(2) Abbreviations for **prefixes of magnitude** (Kilo-, milli, etc) should be written as upper case (capital) letters when greater than 10, and lower case when less than 10. Thus – kilometre as Km, millimetre as mm, hectare as Ha, meganewton as MN.

(3) The prefix and the unit combine to form **one word**, and the abbreviation should be likewise; thus mm for millimetre, NOT m m, m.m or m/m.

(4) **Compound Units** formed by multiplying other units need special care. It is recommended that the abbreviation be written either as N·m or N.m for Newton metre, Kg·m or Kg.m for kilogram metre etc, NOT Nm or N m. Compound units formed by **division** should be written with the sloping bar, e.g. Kg/m. The alternative, $Kg.m^{-1}$, is acceptable, but not recommended for lay users.

(5) The choice of multiple or submultiple to be used is entirely a matter of convenience, although it is usual to choose that which brings the numerical value between 0·1 and 1000. Thus 5.96mm is preferable to 0.005 96m, and 3.61 KN rather than 3610 N. (The second form in each case may be preferable if all other figures in a calculation are in the same units – e.g. metres in the first instance and Newton in the second.) In the special case of engineering drawings dimensions are *always* given in millimetres, no matter how large.

(6) When a unit with prefix is raised to a **power** the power refers to the whole. For example, mm^2 indicates a square millimetre, NOT a 'milli-square metre'. However, with a compound unit the power refers only to the associated unit; thus, KJ/m^3 indicates kiloJoule/ cubic metre. (It is probably wiser for lay users to avoid such powers unless they are accustomed to them, and write cu.mm or sq.m rather than mm^3 or m^2, etc.)

(7) As is common practice in all other systems, numbers having more than four digits (including zero) should be spaced every three digits, thus: 395 561 or 0.000 753 34. The older practice of using commas (e.g. 395,561) is not favoured, as the comma is used as a decimal point in many countries.

None of these rules present any difficulty, and many are common practice. However, it is as well to adhere to them closely to avoid confusion. It is important, too, to be **consistent**; e.g. do not write Kg/m^3 in one place and Kg/cu.m. in another!

One final point here. Abbreviations and mathematical forms are quite acceptable in writing, but in speech they should be avoided. To say 'a speed of fifteen Em-Sec-to-the-minus-one' when you mean 'a speed of fifteen metres per second' is not only clumsy and waste of words, it is a nonsense. There is no such **thing** as a 'second-to-the-minus-one'! $m.sec^{-1}$ is mere mathematical shorthand, not appropriate to a lecture or discussion.

Familiarisation. As my example on page 13 shows, the S.I. system has considerable advantages and these apply especially to those who are not too happy with their 'sums' – calculation is made a great deal easier. However, it must be confessed that, just as my grandchildren have the utmost difficulty with such figures as 13/32 inch (to say nothing of 27/64!) and cannot visualise an ounce or a stone, so those of 'mature years' find S.I. units difficult.

There are two main problems. The first is one of 'scale'. You may recall the apocryphal tale of the aged fitter being introduced to the micrometer. "And how many thous are there **in** an inch. Bert?" "Tiny little things – there must be millions of them" was the reply! A millimetre is pretty small when you have spent a lifetime with inches. (And, of course, our continental friends have the opposite problem when trying to read some of our drawings – it cuts both ways.) However, most of us are now aware that a kilogram is 2.2lb (or 35 ounces), that there are 4½ litres to the gallon, that a millimetre is almost 40 'thou', and so on. Other scale 'models' are that a sq.metre is 10¾ sq.ft., and a Hectare 2½ acres; a cubic yard 35 cubic foot; and so on. In fact, so far as everyday dimensions are concerned normal day-to-day living is accustoming we oldsters to the new regime, and continental travel has shown us that Km/h are as near as makes no odds 5/8 mph!

The main problem lies in the units associated with **force**; new names as well as new sizes; and the 'old' metric users have the same problem. A Newton is 0.225 Lb.force (Lbf) and 1 Lbf is 4.45 Newton in practical terms, not too difficult to keep in mind. (1 Tonf is very nearly 10 kiloNewton.) However, when we come to **pressure** and **stress** it must be admitted that there are difficulties. As shown on page 21 the unit of pressure is the Newton/sq.metre – N/m^2; this is necessary to keep the system consistent, but the result is that 'unit pressure' is very small indeed – 1 Newton is about 3 ounces/sq.yd or 0.000 145 Lbf/sq.in. (almost exactly one grain/sq.in, and the 'grain' started life as the weight of a seed of wheat!). To make matters worse, this unit has a new name – the PASCAL (Pa) – but, in fact, we find that it does not take long to become accustomed to it. It should be noted that it is quite acceptable to state pressure either in Pascals (or its multiples) **or** as a compound unit, Newton/sq.metre. The latter reminds us that we *are* dealing with pressure.

The small magnitude of the Pascal is overcome in one of two ways. The first is to use the MEGAPASCAL (MPa) as the basic unit in practice. This is MN/m^2, and as there are one million sq.mm to the sq.metre, 1MPa is also 1 Newton/sq.mm, or 1 N/mm^2 – rather more practical, as in most applications dimensions will be in millimetres. 1 MPa is about 145 Lbf/sq.in., and 1 Tonf/sq.in. is about 15½MPa. This is not too bad a 'scale'.

The alternative is to use the BAR as the unit of pressure. This is not an S.I. unit, but is commonly used, both here and abroad, when dealing with steam pressures (and, of course, on the Barometer, which is now calibrated in millibar). The Bar is popular on the Continent and in other Metric countries because it is derived from the old CGS system, being 1 Megadyne/cm^2, and has the same order of magnitude as the Kgf/cm^2. The Bar is 0.1 MPa, so that one Bar approximates to atmospheric pressure. However, care must be taken in calculations, as the Bar, being 100 000 N/m^2, does not bear a kilodecimal relationship with the basic units. It is recommended that, while the Bar may be used 'colloquially' when dealing with steam plant, all calculations should be carried out in MPa. However, the MPa or MN/mm^2 should be used when dealing with stresses in materials; the Bar is quite inappropriate here.

The other unfamiliar unit is that for Energy, the JOULE. However, this is a very old unit, and is unlikely to cause much difficulty. It is easy to remember that 1 Joule/sec = 1 Watt (power) and that 1 BThU = 1055 J when dealing with heat. Not quite so easy, perhaps, in mechanical units, 1 Ft.Lbf being 1.356 J in round figures.

The second problem – that of 'familiarity' – is not confined to the S.I. system. Even dyed-in-the-wool 'Imperialists' can be lost with their own system! The

Mechanical Engineer had the same difficulty with Perches as the Surveyor had with micro-inches, and both were lost with the Astronomer's Parsecs! The answer to this difficulty is simple – *use* the system, at first in parallel with the Imperial and later, when the S.I. units become universal, instead of it. The Appendix on page 94 lists some common data in S.I. units. However, I will admit that there is one area where both Metric and Imperial users run into difficulties – that of converting dimensions on drawings, one way or the other. Converting the **figures** from metric to inch or vice versa is easy – you can either use tables or (better) calculate from the conversion factors. But there is rather more to the process than that, and I deal with it in the next section.

Finally, just a word about the arithmetic. The almost complete absence of any conversion factors between units reduces the risk of error considerably, but there remains the hazard of getting the decimal point in the wrong place, especially when using a mixture of 'megas', 'kilos' and 'millis'. The risk can be reduced by using the conventional 'powers to ten' approach, where 1000 is 10^3, 0.000 001 is 10^{-6}, and so on. (These powers are related to the named prefixes in Table 1, page 16.) Then 1987 becomes 1.987×10^3 and 0.004 592 is written as 4.592×10^{-3}. Note, however, that in the field of engineering we always use powers of 3, 6, 9, etc, going up or down in threes. We write 14 563 as 14.563×10^3, **NOT** as 1.4553×10^4. In doing the calculation all the figures are reduced to numbers of S.I. units without prefixes, and multipliers of powers of 10; 1200 MJ being written as 1.2×10^9 J (1.2×10^3 for the 'number' and 10^6 for the 'mega'). A figure of 1200 mm would, of course, automatically become 1.2 metres. The following simple example will illustrate the procedure.

Example. *A gas engine develops 50 KW and uses 14.58 m³ of gas per hour. What is the efficiency if* *the heating value of the gas is 39MJ/m³?*

OUTPUT. 50 KW = 50 KJ/Sec = $\underline{50 \times 10^3}$ J/Sec

INPUT. Gas consumption = 14.58/3600 m³/sec
$$= 4.05 \times 10^{-3} \text{ m}^3/\text{sec.}$$

Hence energy input = $4.05 \times 10^{-3} \times 39 \times 10^6$ J/Sec
$$= \underline{157.95 \times 10^3} \text{ J/Sec.}$$

Efficiency $= \dfrac{\text{Energy Output}}{\text{Energy Input}} = \dfrac{50 \times 10^3}{157.95 \times 10^3}$

$$= \underline{\mathbf{0.317} \text{ or } \mathbf{31.7\%}}$$

You will see that only the **numbers** need be entered on your calculator; the powers of 10 are merely added or substracted as the case may be. Some types of calculator will work in this convention automatically.

The calculation is almost absurdly simple in this case, largely due to the adoption of a 'coherent unit' for energy, the Joule. And here I can, perhaps, make a suggestion to those who feel that the effort to change from Imperial to S.I. units is too great for them (perhaps with the rider 'at my age'!). If all measurements are made in S.I. units in the first place, then there is, quite frankly, no problem at all with calculations. Heating values of fuels, the steam tables, and other required data are all, nowadays, quoted in S.I. units anyway, so that you have to do a conversion if you **must** work in Imperial units! The one major limitation is where a machine tool is calibrated in one system and the drawings are dimensioned in another. I deal with this in the next section, but would remind you that 'conversion indexes' reading in **both** systems are available for many machine tools.

IMPORTANT NOTE. The unit of mass is the **kilogram**, so that 5083 Kg is written as 5.083×10^3 Kg, NOT as 5.083×10^6 gram. Masses measured in grams *must* be brought to kilogram in all calculations; e.g. 38 g = 38×10^{-3} Kg (or, if more convenient, as 0.038 Kg).

Translating Engineering Drawings

The translation of a Metric drawing to Imperial dimensions presents little difficulty, as the amended dimensions would be in decimal inches, and these would imply the same degree of precision as on the original; thus 26mm could translate as 1", or as 1.02" or 1.025" depending on the class of work. Translating from inches to mm differs, however, as many designers still use fractional dimension rather than decimals, and some even work in 'sixty-fourths'. If such fractional dimensions are toleranced the degree of precision is indicated. Even so, the fraction itself may imply working in tenths of thous – 13/16" ± 0.001" means 0.8115" to 0.8135", which converts to 20.612mm to 20.663mm. In the same situation a Metric designer would probably write 20.50 to 20.55mm – a tolerance of ± 0.025mm. The situation is aggravated in drawings of any age, often untoleranced, and even more so on drawings prepared for modelmakers, where the **user** is expected to relate the dimensions of mating parts and so establish the degree of precision required.

The first step is to go systematically over the whole drawing, converting every dimension to the opposite system. I recommend that figures in inches be taken to 4 places of decimals, or if converting to Metric, to three; these can be refined later. Next, seek out and identify the two opposite extremes of precision needed. **First**, those which must tally with a bought-in component – e.g. a ballrace or the bore of a gearwheel which you cannot alter. Mark these plainly and, I suggest, make a note of them. **Second**, mark up figures which are clearly 'rule dimensions'; usually unmachined surfaces, spacing of holding-down bolt-holes and so on.

It is then wise to lay a rule over the castings to assess the amount of machining allowance you have to play with. Note this, in **both** systems of measurement. You should then identify 'matching dimensions' of the less critical kind where, within the limits of the machining allowance, you can alter the dimension of both mating parts to a round number (e.g. a 15/16″ dia spigot (23.81mm) could be probably be made 24 mm – or even 25mm – with no problem). The point here, as with the 'ruler' dimensions, is to select conversions which are easy to measure in the opposite system, whether by calipers or with the micrometer or vernier.

The next step is more important. That is to establish reference faces and centrelines so that working parts will align. The point is easily appreciated; dimensions in fractional inches will seldom convert to round hundredths of millimetres – the smallest increment that you can work with a normal micrometer. The throw of a crankshaft, for example, or the location of an eccentric relative to the valve-chest, must be dimensioned to an exact match. This must be done carefully and, I suggest, the revised dimensions entered on the drawing as you proceed. Having checked them you can then fill in any remaining dimensions – it is quite certain that some of those previously determined will have to be altered.

My own practice at this stage is to make a new drawing altogether, filling in the Metric dimensions (or inch, if I am converting **from** Metric) as I would were I doing the job from scratch. But there is, in these days of accessible copiers, a very useful aid. Make a copy of the original drawing and go over this deleting all the dimensions with 'Liquid Paper' – the special sort for photocopies. Then get copies made of this undimensioned drawing. You can then enter on one copy the literal conversions you made at the beginning, as well as the reference faces and centrelines. The second copy can then be used for the various steps mentioned, and the third used for entering your final decisions.

I have shown the process on a relatively simple example. This (Fig.1) is the crankcase, crankshaft and bearing housing for a small two-cylinder single acting high-speed steam engine for a model boat. I have not shown all the details as by the time the drawing is reduced to book size it would have been smothered in figures. (Most of the omitted detail is concerned with stud positions, oil-holes and the like, which present few problems). Fig.1 is the original 'inch' drawing and you will see that I have added 'reference' details; the centrelines of the block and crank, and the length (½ inch) of the boss on the right-hand bearing housing, which locates the crank endways. It is, of course, obvious that the spacing of the two crankpins must align with the spacing of the cylinder bores. I have also added machining marks – often absent from drawings published for use by modelmakers, unfortunately.

Before going further there is one decision which had to be made – that of the cylinder bore (and, perhaps, the stroke). A Metric design of engine would have a bore of 25mm, not 25.4mm, and if 25mm piston rings were available this is the proper conversion. In this case, however, the piston rings were to hand for a bore of 1.000 inch; to reduce the bore to 25mm would mean filing at the gap, removing 1.26mm (about 0.050″) and though this presents no difficulty there would be a risk that the ring might not fit properly. So, the decision is made to keep to 1″ or 25.4mm. And as there is no difficulty with the crank either, the throw is retained at the metric equivalent of ½ inch – 12.7mm.

Fig. 2 shows a second copy of the drawing dimensioned with **exact** metric conversions, though all have been rounded off to the third decimal place. You will notice a number of dimensions marked with an asterisk – *; these are all 'rule dimensions', which can be altered within the limits of the machining allowance (3/32 inch or 2.4mm) to suit the machinist's convenience.

We have one dimension (apart from the cylinder

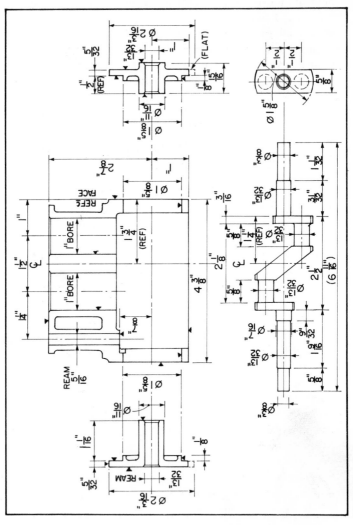

Figure 1

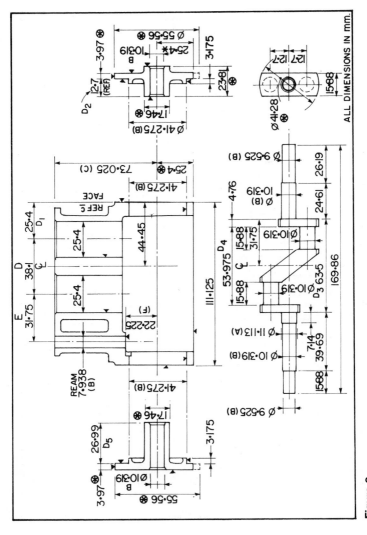

Figure 2

bore) which must fit a bought-in part; the 7/16 in. dia
on the shaft, marked 'A', carries a commercial bevel
gear. Compare Fig.2 with Fig.3, and you will see that
it is dimensioned at 11.11mm (0.437 in) The machinist
can work to either, the difference between the two
being 0.003mm or 0.0.0001″, not significant.

There are some places where parts can be machined
to a mutual fit, marked 'B'. The end bearing housings,
both the spigot and the bearing bore; the mating part
of the shaft; the ends of the shaft, which fit to flywheel
and coupling; and the vertical reamed hole which
carries the bevel-wheel shaft. Consider these in turn.
The spigot can be 42mm, not critical provided the crank
can be passed through. The shaft bearing can be 10mm,
a standard reamer size; a little homework shows that
the combined bending and torsion stress will still be
reasonable. The shaft ends can be 9mm with no harm,
still standard reamer size, and the vertical shaft 8mm.
Now consider the dimension 'C' – the height of the
block above the crank centreline. This dimension
would affect the piston clearance at dead centre.
Examination of the connecting rod forging showed that
centres had to be held fairly close. The exact 73.025 is,
therefore, held at 73mm, instead of being altered to 75
which would be a more 'normal' dimension here.

Now for the critical part – aligning crank and block.
Compare Figs.2 and 3 during the next stages. The
cylinder centre distance, 'D' can be refined to 38mm,
and that from the centre of bore No.1 to the reference
face made 26mm without over-running the machining
allowance – D_1. (We increase D_1 as a little, as we
decreased 'D'.) This brings the **reference dimension**,
originally 44.45mm, to 45mm. Again, a convenient
round figure. Turn to the crankshaft. The **outer** faces
of the two webs determine the shaft location, 'D_3'
Make this 64mm for trial. Then, if the crankpins are
made 16mm wide (the big end bearing can be machined
to suit) and the webs made 5mm thick the centre
distance of the two crankpins works out at 38mm,

exactly as required for the cylinder bores. It only remains to adjust the length of the location boss on the R.H. end plate to match up (D_2) and this works out at 13mm. To summarise, we find D at 38, D_1 at 26, D_2 at 13, D_3 at 64, and crankpin width at 16mm. The only remaining dimension of this group is D_5, the boss on the left-hand end cover. This has been set at 28mm (not 27) to allow the overall length of the block to be rounded to 112mm. Some adjustment will be needed at erection in any case, to allow for setting the meshing of the bevel gears.

Apart from details of 'rule' dimensions the only remaining to be settled are, first, the crankpin diameter and secondly, the dimensions marked 'E' and 'F'. The crankpins can be either 10mm or 11mm, both standard reamer sizes. If 10mm, the stress will go up by about 9% – well within the safe limit for drop-forged steel – and I would use that figure, if only to avoid yet another reamer size. However, those who have an eye to increasing bearing life would certainly plump for 11mm dia here. The lengths of the various steps in the shaft are not at all critical, and would be simply rounded from the 'exact' figures.

Now for 'E' and 'F'. The figure at 'E' can affect the valve timing, (the engine has a transverse piston valve operated by a link from the vertical shaft). Slight adjustment is possible. The **exact** dimension from the block centreline is 2 inches = 50.8mm. If 'E' is made 32mm, this dimension becomes 51mm, a difference of 0.2mm, about 0.008 inch. But we have, as it happens, moved the seating of the bevel gear on the shaft endways by 0.25mm, which compensates almost exactly. So, 'E' can be 32mm with no problems. Finally, 'F'. This figure affects the meshing of the two bevels. The distance is, therefore, rounded **upwards**, so that shim-washers can be used for adjustment. In all this work it will be helpful to remember that 0.1mm is almost four thou (0.004″) and 0.01mm four tenths. And, for the Metric user, one thou (0.001″) is 0.025mm

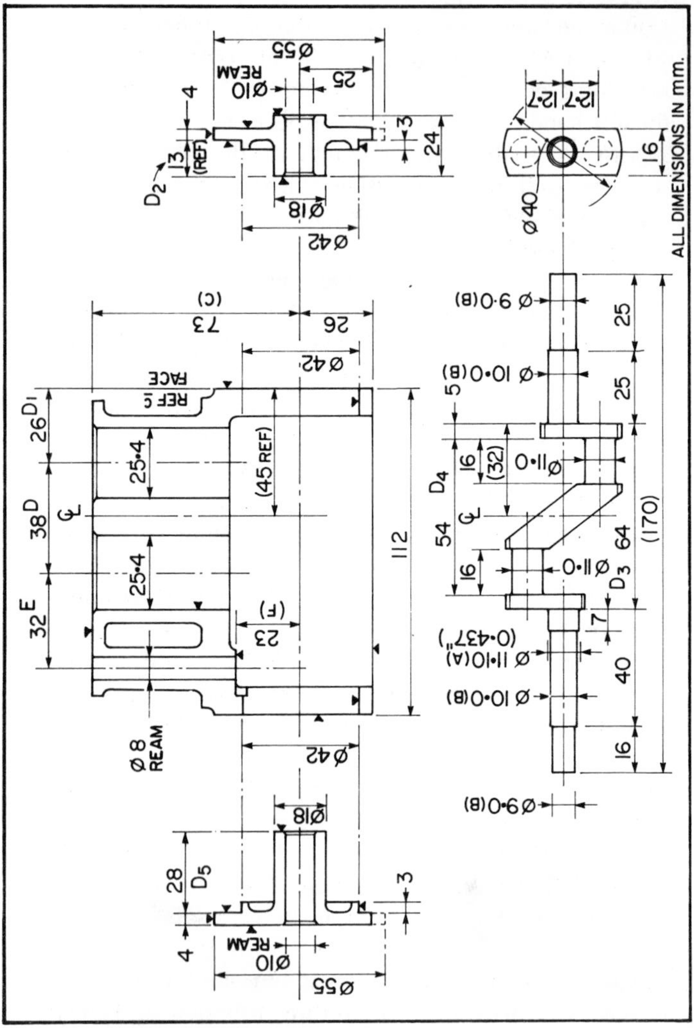

Figure 3

when converting drawings in the opposite direction.

References between Figs. 2 and 3 will show how the other, minor, dimensions have been settled and you will notice that we have been able to make the final conversion in round figures of millimetres with very few exceptions. We have, in effect, redesigned the four components to S.I. standards.

Agreed, this exercise has taken some time, and to deal with all the lesser dimensions not shown on the sketches – **and** those on the two or three score other parts – would take even longer. As I said before, we are really redesigning the engine with the constraint of having to work to the existing castings, and this, inevitably, takes longer than designing an engine and leaving the materials supplier to make patterns and castings for you!

On a really large set of drawings, of course, the job would be formidable indeed, and some would say that it was impossible. This is far from being the case. It is, in fact **easier** than reducing the prototype from full size to model scale; at 1 in/ft 19 inches comes out at 1.583333 ins! However, it does take time, and I find that by far the best procedure when a number of really large drawings are involved is, after carrying out the initial steps (identifying rule, mating, and 'exact' dimensions, and then making an exact conversion of all) to get out the drawing-board and make a new set of drawings altogether. I would add a further piece of advice, too. If scaling down a full-sized prototype but to Metric dimensions it is **far** easier (and safer) to use a 'decimal' scale – 1/10 instead of 1/12. This cannot be done for locos which must run on an 'Imperial' track, of course, but for other model work there are very considerable advantages.

I need not add that in all cases it is vital to check and double-check your conversion drawings! I might add that such checks may often reveal some errors in the original drawings **before** conversion!

Preferred Sizes of Stock Materials

Tables 18 to 22 show the present (1991) sizes for bar, wire, sheet, tube drills and reamers, and timber as held by normal local stockists, though wider choice may be expected from larger merchants. The 'R' series of preferred numbers has not, so far, been rigidly observed in the UK, where stockists appear to be offering a compromise between the 'R' series and the former 'Inch' preferred values. (Continental readers will appreciate that UK practice was to size barstock in **fractions** rather than decimals.) In the final analysis 'preferred' sizes will always emerge as those most convenient to users, and enquiry at the local stockist may often produce non-listed stock. Table 22 page 92 presents the I.S.O. metric screw thread range. Whitworth, Unified, B.S.Pipe and B.A. threaded fasteners are still freely available, though the hexagon sizes on BA screws are now almost entirely in 'near BA' Metric sizes.

DRILLS & REAMERS

Table 16 presents recommended alternatives to the former 'Number & Letter' series, and these are the sizes which will be supplied if a number drill is ordered. No table of preferred sizes of either drills or reamers is included. Preferred drills run from 1mm to 13mm × 0.1mm in 'jobbers' style (straight shank) and from 12 to 25mm × 0.5mm in No.2 MT taper shank.

Hand reamers are available in steps of 0.5mm from 1.5 up to 15mm dia, and then in steps of 1mm up to 25mm dia. though not all are 'ex-stock'.

Table 18 Drawn & Rolled Rod or Bar

Rounds Both ferrous and non-ferrous barstock is manufactured in steps of 0.5mm from 3mm up to 20mm, and thereafter in 1 mm steps. However, many sizes are available only in tonnage or ex-mill quantities. Typical 'Stockists' Preferred Sizes' are:

    3mm,   4mm,   5mm,   6mm,   8mm, 10mm,
    12mm, 14mm, 15mm, 16mm, 18mm, 20mm,
    25mm, 30mm, 35mm, 40mm, 50mm, 60mm,
    70mm, 80mm, 90mm, 100mm

Hexagons Sizes are limited to those required for the I.S.O. screw threads, viz:

    4.0mm,   5.0mm,   5.5mm,   7.0mm,   8.0mm,
    10.0mm, 13,0mm 17.0mm, 19.0mm, 22.0mm,
    24.0mm, 24.0mm, 27.0mm, 30.0mm, 32.0mm 36.0mm
    All across flats.

Small A–F hexagons to suit Socket-head screws are as follows, but not freely available except in high-tensile steel at present.

    0.71mm. 0.89mm, 1.27mm, 1.5mm,
    2.00mm, 2.50mm, 3.00mm, and then as above.

Squares are nominally available in the same sizes as rounds, but stockists preferred sizes are more limited, typically:

    3mm,   4mm,   5mm,   6mm
    8mm, 10mm, 12mm,
    15mm, 20mm, 25mm, 30mm.

Table 19 Wire and Sheet

Metric sizes of both wire and sheet approximate to the R20 number series (see page 70) but wire in SWG (Standard Wire Gauge) is still (1991) widely used, especially in the motor rewinding trade.

Wire

Dia, mm	App. SWG	Dia, mm	App. SWG	Dia, mm	App. SWG
0.100	42	0.630	23	2.5	12
0.125	40	0.710	22	2.80	11
0.160	38	0.80	21	3.15	10
0.200	35	0.90	20	3.55	9
0.250	33/34	1.00	19	4.0	8
0.280	31/32	1.12	–	4.5*	7
0.315	30	1.25	18	5.0*	6
0.355	28/29	1.4	17	5.6*	5
0.400	28	1.6	16	6.3*	3
0.450	26	1.8	15	7.1*	2
0.500	25	2.0	14	8.0*	1
0.560	24	2.24	13		

* Usually classed as 'ROD', included for SWG comparison only

Sheet

Copper and Brass Thickness follows the R20 series in mm.

0.02	0.060	0,20	0.70	2.2
0.022	0.070	0.22	0.80	2.5
0.025	0.080	0.25	0.90	2.8
0.028	0.090	0.30	1.0	3.0
0.030	0.100	0.35	1.1	3.5
0.035	0.11	0.40	1.2	4.0
0.040	0.12	0.45	1.4	4.5
0.045	0.14	0.50	1.6	5.0
0.050	0.16	0.55	1.8	6.0
0.055	0.18	0.60	2.0	

Aluminium. Thicknesses do *not* follow the R-series of numbers.

Thin sheet ('Foil') runs 0.005mm to 0.01mm × 0.001mm, then 0.012, 0.015, 0.018 0.020, 0.023: 0.025 to 0.050 × 0.005: 0.06, 0.065, 0.075, 0.090: Then 0.100 to 0.175 × 0.025 steps.

'Sheet' runs from 0.25mm to 3.0mm, × 0.25: then 0.3 to 1.0 × 0.1 steps: finally 1.2, 1.6, 1.8, 2mm, 2,5mm and 3mm.

Material thicker than 3mm is listed as **'Plate'** running from 3mm to 15mm × 1mm, and from 15 to 50mm × 5mm, though smaller increments are available from some sources.

Steel. Thicknesses from 0.3mm (say 30SWG) upwards follow those for copper up to 2 to 3mm, after which the material is classed as 'Strip', thickness rising from 2 to 6mm × 1mm steps. Above 6mm is classed as 'Plate', preferred thicknesses being 10, 12, 15, 16, 20, 25, 30, 40 and 50mm upwards.

Table 20 Copper Tubes

In the table below columns **X, Y,** and **Z** refer to tubes for capillary jointed domestic water, gas and sanitation systems. These have a close tolerance (from ± 0.04 to ± 0.1mm, depending on size) on the O.D. The columns marked 'Other Sizes' gives the thicknesses available in the 'Pipeline', 'Steam Services' or 'General Purposes' I.S.O. specifications. The table lists the most common sizes; others will be found in BS 2871, including a series having nominal "inch" bores (1/8 in up to 4 in) but with metric thickness, for threaded services.

NOM O.D.	Thickness, Capillary "X"	"Y"	"Z"	Other Thicknesses		
3	–	–	–	0.5	0.6	0.8 ***
4	–	–	–	0.5	0.6	0.8
6	0.6	0.8	0.5	1.0		
8	0.6	0.8	0.5	1.0		
10	0.6	0.8	0.5	1.0	0.7	
15	0.7	1.0	0.5	1.5		
18	0.8	1.0	0.6			
22	0.9	1.2	0.6			
25	–	–	–	1.0	1.5	2.0
28	0.9	1.2	0.6	1.5		
30	–	–	–	1.5	2.0	
35	1.2	1.5	0.7	2.0		
42	1.2	1.5	0.8	2.0		
50	–	–	–	1.5	2.0	2.5
54	1.2	2.0	0.9	3.0		
67	1.2	2.0	1.0	3.0		
76	1.5	2.0	1.2	2.5	4.0	
108	1.5	2.5	1.2	3.0	3.5	4.9
133	1.5	3.0	1.5	6.5		
159	2.0	3.0	1.5	2.5	3.5	8.0

*** Small bore copper tube is still (1991) widely available in SWG thicknesses. Small-bore tube below about 1.5mm O.D. is known as 'Capillary Tubing'

Table 21 Timber

Standard Sizes for Sawn Softwood

Thickness mm	Width – Millimetres						
	75	100	125	150	175	200	225
16	√	√	√	√			
19	√	√	√	√			
22	√	√	√	√			
25	√	√	√	√	√	√	√
32	√	√	√	√	√	√	√
38	√	√	√	√	√	√	√
44	√	√	√	√	√	√	√
50	√	√	√	√	√	√	√
63		√	√	√	√	√	√
75		√	√	√	√	√	√
100		√		√		√	

STANDARD LENGTHS 1·8m to 6·3m × 0·3m intervals.

Note Timber ordered "PAR" (Planed all round) will be slightly smaller – thus 100 × 25 PAR will have been planed down *from* 100 × 25 sawn section: If ordered "PAR to dimension 100 × 25" the timber must be resawn from larger size and will be more expensive.

I.S.O. Metric Screw Threads

There are two series of fasteners – 'Normal' or coarse thread, and 'Fine' thread. The former are denoted by the prefix 'M' and diameter – e.g. M5; fine threads by the same prefix, but followed by the pitch – M5 × 0.5. In addition, there are five constant pitch series, 0.75mm, 1.0mm, 1.5mm, 2mm and 3mm pitch, and one 'special', M14 × 1.25 for sparking plugs. The series runs up to 300mm dia × 6mm pitch. Table 22 gives 1st, 2nd & 3rd preferences up to 56mm diameter.

A series of 'Instrument Threads' runs from 0.3mm dia × 0.08mm pitch up to 1.4mm × 0.3. These are designated by the prefix 'S', thus S–0.6.

A further approved (but non-S.I.) metric series specifically for model engineers is specified in BSI PD.6507 of 1982. These have hexagon sizes which are more 'to scale' and, in addition, offer 'constant pitch' sizes which approximate to the Imperial (Whitworth form) used in model engineering. Full details are given in *The Model Engineer's Handbook* and in *Drills, Taps & Dies* both from Argus Books. The latter gives full details of the thread forms, stress areas and tapping drill sizes.

Important Note: Although the 'miniature' S series has the same thread angle as the M series the crest truncation is different. This means that male and female components having the same diameter and pitch in the two series may not match.

Table 22
M Series

1st Pref Dia	1.0		1.2		1.6		2.0		2.5	3.0		4.0		5.0		6.0		8.0		10.0	
2nd Pref Dia		1.1		1.4		1.8		2.2			3.5		4.5						9.0		11.0
3rd Pref Dia															5.5		7.0				
Normal Pitch	0.25	0.25	0.25	0.3	0.35	0.35	0.40	0.45	0.45	0.5	0.6	0.7	0.75	0.8	–	1.0	1.0	1.25	1.25	1.5	1.5
Fine Pitch	0.20	0.20	0.20	0.20	0.20	0.20	0.25	0.25	0.35	0.35	0.35	0.5	0.5	0.5	0.5	0.75	0.75	1.0	–	1.25	–
Const 0.75	#	#	#	#	#	#							0.75			0.75	0.75	0.75	0.75	0.75	
Const 1.0	¶	¶	¶				¶	¶								1.0	1.0	1.0	1.0	1.0	1.0
Const 1.5										*		*	*	*	*						1.5
Const 2.0																					
Const 3.0																					

#, Constant pitch of 0.2mm. ¶, constant pitch of 0.25mm. *, constant pitch of 0.5mm.

Table continued on page 93.

Table 22 continued

	12	14	15	16	17	18	20	22	24	25	26	27	28	30	32	33	35	36	38	39	42	48	56
1st Pref Dia	12			16			20		24					30				36			42	48	56
2nd Pref Dia		14				18		22				27				33				39			
3rd Pref Dia			15		17					25	26		28		32		35		38				
Normal Pitch	1.75	2.0	–	2.0	–	2.5	2.5	2.5	3.0	–	–	3.0	–	3.5	–	3.5	–	4.0	–	4.0	4.5	5.0	5.5
Fine Pitch	1.25	1.5	–	1.5	–	1.5	1.5	1.5	2.0	–	–	2.0	–	2.0	–	2.0	–	3.0	–	3.0	4.0	4.0	4.0
Const 0.75																							
Const 1.0		1.0	1.0	1.0	1.0	1.0	1.0	1.0	1.0	1.0		1.0	1.0	1.0									
Const 1.5		1.5	1.5	1.5	1.5	1.5	1.5	1.5	1.5	1.5	1.5	1.5	1.5	1.5	1.5	1.5	1.5	1.5	1.5	1.5	1.5	1.5	
Const 2.0						2.0	2.0	2.0	2.0	2.0			2.0		2.0			2.0		2.0	2.0	2.0	2.0
Const 3.0									3.0					3.0		3.0		3.0		3.0	3.0	3.0	3.0

"S" Series – Miniature Screw Threads

1st Pref Dia	0.3		0.4		0.5		0.6		0.8		1.0			1.4*
2nd Pref Dia		0.35		0.45		0.55		0.7		0.9		1.1*	1.2*	
Pitch	0.08	0.09	0.10	0.10	0.125	0.125	0.15	0.175	0.20	0.225	0.25	0.25	0.25	0.3

* Not Recommended – Use M Series.

Appendix

The data needed for most engineering calculations is, nowadays, quoted in S.I. units in the relevant textbooks. However, many readers may have access only to books using either the Imperial or the MKS systems. The data can, of course, be converted using the tables given earlier, but for convenience, some of the more commonly needed figures are given below.

(1) *Average Properties of Materials*

	U.T.S. N/mm^2	Yield Point N/mm^2	Young's Modulus (E) KN/mm^2	Notes
Grey Cast Iron	150–180	(620)*	82–104	()* Compressive Strength
Malleable Iron	310–390	120–150	76	
FCMS 220M07	430–460	310	207	En1A
0.2%C 070M20	525	370	207	EN3
0.4%C 080M40	540	280	207	EN8
Stainless 302S25	780	550	193	EN58A
Silver Steel	610–920	540–770	207	
Copper (	210	46	110	Annealed
(	260	215	110	Half-hard
Brass (	340	215	103	70/30 drawn rod
(	410	215	96	60/40 drawn rod
Gunmetal, Cast	260–300	120–185	110	
Phos. Bronze	510	390	100	Drawn Rod
Monel "400"	510–680	200–370	172	
Y-alloy LM14	155–215	125–155	70	drawn Rod
RR56 alloy	400	310	72	forged

(2) *Average Properties of Fuels*

(a) *Solids*

	Sp. Energy/ unit mass	Air Required	Energy release/ unit vol. air.
	MJ/Kg	m³/Kg	MJ/m³
Soft Coal	25.78	6.7	3.85
Steam Coal	33.66	8.7	3.87
Anthracite	34.38	8.8	3.91
Coke	30.21	7.9	3.82

(b) *Liquids*

Petrol	47.12	11.5	4.10
Kerosene	46.66	11.3	4.13
"Derv"	45.9	11.09	4.14
Alcohol (Meths)	23.69	6.16	3.85

(c) *Gases*

	Sp. Energy/ MJ/m^3	Air req'd Vol/vol m^3/m^3	Energy release/ unit volume of air-fuel mixture MJ/m^3
Methane CH_4	35.57	9.52	3.74
Butane C_4H_{10}	118.66	30.0	3.96
Propane C_3H_8	92.7	23.0	4.03

(3) *Properties of air*

Specific Heats: C_p = 1.005 KJ/Kg°C. C_v = 0.718 KJ/Kg°C. "Gamma" (γ) = 1.4.
Specific volume at std. atmos and 0°C = 0.774 Cu.m/Kg.
Specific gas constant, R = 0.287 KJ/Kg°C. Mean molecular weight = 28.96.
Standard Atmospheric pressure = 101.325 KPa = 1.01325 Bar

(4) *Properties of steam* (Skeleton Table)

Abs. Press. Bar	Temp. °C	Specific Volume m³/Kg	water KJ/Kg	Enthalpy latent KJ/Kg	steam KJ/Kg	Abs. Press Lbf/sq.in
1.0	99.6	1.694	417	2258	2675	14.50
1.5	111.4	1.159	467	2226	2693	21.8
2.5	127.4	0.7186	535	2182	2717	36.3
3.0	133.5	0.6057	561	2164	2725	43.5
4.0	143.6	0.4623	584	2148	2732	58.0
5.0	151.8	0.3748	640	2109	2749	72.5
6.0	158.8	0.3156	670	2087	2757	87.0
7.0	165.0	0.2728	697	2067	2764	101.5
8.0	170.4	0.2403	721	2048	2769	116.0
9.0	175.4	0.2149	743	2031	2774	130.5
10.0	179.9	0.1944	763	2015	2778	145.0
11.0	184.1	0.1774	781	2000	2781	159.5
12.0	188.0	0.1632	798	1986	2784	174.0

Approximate specific heat of superheated steam:—

At 7 Bar, 2.12 MJ/Kg°C

At 10 Bar, 2.20 MJ/Kg°C